RESOURCING THE NATIONAL SECURITY ENTERPRISE

RESOURCING THE NATIONAL SECURITY ENTERPRISE

EDITED BY

Susan Bryant and Mark Troutman

Rapid Communications in Conflict and Security Series
General Editor: Geoffrey R.H. Burn

CAMBRIA
PRESS

Amherst, New York

Requests for permission should be directed to:
permissions@cambriapress.com, or mailed to:
Cambria Press
University Corporate Centre, 100 Corporate Parkway, Suite 128
Amherst, New York 14226, U.S.A.

Library of Congress Cataloging-in-Publication Data

Names: Bryant, Susan F., editor. | Troutman, Mark, editor.

Title: Resourcing the national security enterprise: connecting the ends and
means of US national security / edited by Susan Bryant and Mark Troutman.

Description: Amherst, New York: Cambria Press, [2022] |
Series: Cambria rapid communications in conflict and security series |
Includes bibliographical references and index. | Summary: "Considering the national
security enterprise from the standpoint of strategic resourcing is neither simple nor
straightforward. To succeed requires a multidisciplinary approach; a group of writers
with substantial background knowledge on such diverse and byzantine topics as the
Department of Defense acquisition system, the President's budget submission and the
Federal Emergency Management Agency's National Preparedness Frameworks; as well
as a basic understanding of macroeconomics. Further, the development of a cohesive and
logical narrative is difficult because the Framers' intended checks and balances among the
executive and legislative branches effectively preclude the possibility of seamless integration
among national security priorities. Each chapter in this book was written by a practitioner
with decades of experience working resourcing issues in Washington. Their perspectives
are informed by the cultures of the agencies in which they have worked and the positions
they have held. Many currently teach in D.C. based graduate degree programs in a variety
of disciplines from strategy to economics to organizational leadership. Thus, this book is
intended as a theoretically grounded yet practical guide for those who seek to understand
the inner workings of the American federal government."-- Provided by publisher.

Identifiers: LCCN 2021049588 (print) | LCCN 2021049589 (ebook) |
ISBN 9781621966227 (library binding) | ISBN 9781621966241 (paperback)
ISBN 9781621966371 (pdf) | ISBN 9781621966388 (epub)

Subjects: LCSH: National security--United States--Finance. | Security sector--
United States--Finance. |United States. Department of State--Appropriations
and expenditures. | United States. Department of Defense--Appropriations
and expenditures. | National security--United States--Decision making.

Classification: LCC UA23. R455 2022 (print) | LCC UA23 (ebook) |
DDC 355/.033073--dc23/eng/20211208

LC record available at https://lccn.loc.gov/2021049588

LC ebooks record available at https://lccn.loc.gov/2021049589

TABLE OF CONTENTS

LIST OF FIGURES

FOREWORD

Alongside international security challenges, the United States has also always faced domestic financial and budgetary challenges. Since its founding, the country's strategic choices have been constrained and shaped by the economic and financial resources at its disposal. This was the case during the Cold War with the former Soviet Union and during America's "unipolar moment" after the collapse of the Soviet Union—and it is still the case today. If anything has changed, it is the difficulty and complexity of the tradeoffs involved as well as the imperative to make them wisely and efficiently.

During the Cold War, the United States faced a security challenge that, while by no means solely military in nature, had a military component that unmistakably represented a "clear and present danger." Fortunately, during that era, the United States and its NATO allies also held a sizable economic advantage that only grew with time. During the early post-

Cold War period, the security challenges facing the United States became less pressing but also more diverse, complex, and multidimensional.

The existential military threat collapsed with the Soviet Union, but serious challenges emerged that were better confronted with economic power than military force. In recent years, the challenges faced by the United States have grown even more multidimensional and complex in nature. Rather than a purely military threat, the challenges facing the United States now encompass issues such as cybersecurity, nuclear proliferation, terrorism, failed states, economic competition, climate change, and a global pandemic—and these are all in addition to the more traditional military and diplomatic challenges, especially in the form of a rising China.

Simultaneously, domestic trends within the United States threaten to constrain US options for spending on national security. The aging population and other long-term demographic changes, in conjunction with rising health care costs, have long posed serious budgetary challenges that will grow in coming years. Between 2020 and 2050, the proportion of the US population over 65 years of age will increase from about 16 percent to 22 percent, while the working-age proportion will decline from 58 percent to 54 percent.[1] Moreover, although often associated with the retirement of the baby boomer generation, this demographic shift represents a permanent change—driven, among other things, by the long-term decline in birth rates common to developed countries.

These long-term challenges have been exacerbated by economic down-turns, a series of large tax cuts, and, most recently, by the global COVID-19 pandemic with its enormous and unprecedented health and economic impacts. Largely as a consequence of the pandemic—including both the economic recession it triggered and the increased spending for relief that it necessitated—the size of America's publicly held federal debt grew by some $4.2 trillion in the fiscal year 2020 alone.[2]

After declining from its World War II peak, the federal debt hovered at between roughly 25 percent and 50 percent of gross domestic product

(GDP) for decades and into the beginning of the twenty-first century. Since the Great Recession, however, the trend has moved upward substantially and consistently, with the level of debt reaching 61 percent of GDP by 2010, 77 percent by 2018, and 98 percent by 2020.[3] Moreover, the situation is projected to get far worse over the next several decades. In its baseline projections—which assume no major changes in revenue or spending policies—the Congressional Budget Office (CBO) projects that by 2050 the federal debt will grow to some 202 percent of GDP.[4] Some other projections forecast less dramatic growth in the debt, and there is room for reasonable minds to differ in terms of how quickly and substantially the growth of the debt needs to be addressed.[5] Regardless, all credible projections show a progressively worsening and challenging situation.

Ultimately, America's national security rests in large measure on the strength of its economy as well as its domestic social and political cohesion. The importance of a strong economy to meeting US national security requirements stems directly from the power and influence that flows from a robust economy at home and intense economic engagement abroad as well as indirectly from the strong economy needed to underwrite military, intelligence, foreign affairs, homeland security, and other national security capabilities. Failure to address the growing federal debt threatens to undermine both of these economic aspects of our national security. Financing a large and growing federal debt will increasingly crowd out the private investment needed to sustain economic growth. Likewise, a large federal debt will make it increasingly difficult for the United States to cushion the impact of economic downturns with counter-cyclical fiscal policies and leave the United States increasingly vulnerable to another potential financial crisis.

Effectively addressing the country's worsening fiscal situation will require more than simply restraining military spending along with reducing other national security capabilities. National security spending —all in—accounts for at most perhaps a quarter of federal spending. Moreover, the largest and fastest growing parts of the federal budget are related to other areas, particularly spending on major medical programs, such as Medicare and Medicaid, and, to a lesser extent, social security. Similarly, revenue increase will almost certainly have to be part of any realistic solution—given how so many Americans are dependent on these and other critical federal programs.

However, while restraint on defense and other national security spending clearly cannot be the whole solution to addressing the growing federal debt, it is hard to imagine some level of restraint not being a part of any solution. Curtailing national security spending has been a central element of every major deficit-reduction effort of the past half century, from the Graham-Rudman-Hollings Balanced Budget Act of the 1980s to the Budget Enforcement Act (BEA) of the 1990s and, most recently, the Budget Control Act (BCA) of 2011. There is every reason to believe it will be a key element in any future effort to bring US spending and revenue back into balance. In short, over the next few decades it is likely to become ever more difficult for the United States to address national security challenges through simply spending more. Increasingly, we will need to spend smarter.

Spending smarter means determining how best to address the full range of future security challenges. When are they best addressed by improving economic strength at home or our investment and trading power abroad? The rise of China has driven home how critical economic growth can be for generating a country's influence abroad not just in economic but also social, political, and diplomatic spheres. During the Cold War, the United States had little to fear from the Soviet Union's appeal or reach as an economic power. China's economy today represents a far more significant challenge to American influence. Does this challenge call for more spending on defense, infrastructure, or other potential triggers for economic growth?

When are diplomacy and foreign assistance the most cost-effective tools? Many observers believe that the past two decades have been marked by a dangerous erosion in the influence of diplomacy and development as well as an overreliance on military capabilities—with results that have been problematic or worse. In many cases, diplomacy and foreign aid might have proven to be more cost-effective tools. But here, too, the choices are not so simple. In a complex world where direct foreign investment, remittances, and private charities play increasingly important roles, determining how best to apply the instruments of diplomacy and foreign aid is no easy task.

What role should homeland security play? After decades of intermittent and uneven attention, the homeland security mission returned as a central national security concern after the terrorist attacks of 9/11. Nevertheless, the place of homeland security in the hierarchy of US national security missions remains uncertain, as does the most cost-effective way to organize and carry out the extraordinarily diverse and complex set of missions that fall under this rubric.

And when and where must we continue to focus on countering traditional (and nontraditional) military threats to our security? Over the past several decades, China has emerged as a potential peer competitor with increasingly advanced military capabilities. But it represents by no means the only potential military challenge. Russia retains a powerful range of military capabilities, especially in the nuclear realm, while countries like Iran and North Korea represent smaller but—in the eyes of some at least—even more difficult challenges. And what about less conventional military missions, such as counterterrorism and stability operations? Even if we agree that military capabilities remain central to US national security requirements, many questions remain concerning tradeoffs among various missions as well as the most cost-effective approach to carrying out each of these missions.

And, of course, to be effective, we must also understand the processes used to develop and execute these policies, plans, programs, and budgets —not only within the executive branch and Congress but also in each of the different areas that comprise our broad national security toolkit. Too often, process is treated as an afterthought, or at best a secondary consideration. In fact, process plays a key role in shaping outcomes in the national security world, as it does elsewhere. The planning process differs in each of the federal departments and agencies charged with carrying out national security missions. But they all share some common elements, including—perhaps most importantly—budgeting. Indeed, to a large degree, across all departments and agencies, the budget process represents the mechanism through which plans, policies, and, most particularly, programs are put in place. And this is no less true within Congress than within the executive branch. Gaining a better understanding of the planning and budgeting processes through which our diverse national security capabilities are shaped is key to improving the effectiveness and cost-effectiveness of those capabilities.

Finally, we must increasingly think about how best to align our budgetary resources with specific national security outputs, like foreign assistance, weapons modernization, military readiness, and force structure plans—all of which constitute complex components of our national security toolkit. Improving our ability to effectively make tradeoffs between and among different types of national security capabilities requires us to have a better understanding of not only the broad tools with which we craft our national security but also the detailed components that comprise those tools.

In this context, the timing and content of this book could not be more opportune. The long-term challenges the United States faces are as demanding as any it has faced over the past half century. On the domestic front, we face unprecedented demographic trends that—when combined with continued growth in per capita healthcare costs, a series of large tax cuts, the need to address climate change, an aging infrastructure, and other priorities—will make it more important than ever to rationalize federal program spending and revenue policies. At the same time, the United States faces a security environment which has perhaps never been more complex or diverse. The chapters in this book do not offer definitive answers to the multitude of questions raised by these challenges, but—

far more valuable—they provide a long-needed introduction on how to think about these questions and how to begin to evaluate the answers to these questions proffered by those inside the very broad and diverse US national security community.

Steve Kosiak is a partner with ISM Strategies, a consulting firm based in Washington, DC. He served for five-and-a-half years as the Associate Director for Defense and International Affairs for the Office of Management and Budget of the White House.

NOTES

1. Jonathan Vespa, Lauren Medina, and David M. Armstrong, "Demographic Turning Points for the United States: Population Projections for 2020 to 2060," Bureau of the Census, (February 2020): 1, 6, https://www.census.gov/content/dam/Census/library/publications/2020/demo/p25-1144.pdf.
2. U.S. Department of the Treasury, "Historical Debt Outstanding—Annual 2000–2020," https://www.treasurydirect.gov/govt/reports/pd/histdebt/histdebt_histo5.htm.
3. Congressional Budget Office, "Federal Debt: A Primer" (March 2020): 2, 4. [https://www.cbo.gov/publication/56309]
4. Congressional Budget Office, "The 2021 Long Term Budget Outlook" (March 4, 2021): 7, https://www.cbo.gov/publication/57038.
5. See, for example, Jason Furman and Lawrence Summers, "A Reconsideration of Fiscal Policy in the Era of Low Interest Rates," November 30, 2020, https://www.piie.com/system/files/documents/furman-summers2020-12-01paper.pdf.

RESOURCING THE NATIONAL SECURITY ENTERPRISE

CHAPTER 1

INTRODUCTION TO RESOURCING NATIONAL SECURITY

Susan Bryant

> I wish I had the sort of job you think I have.
> —Allen Greenberg,
> *Confessions of A Government Man*[1]

INTRODUCTION

This volume is the book that we, the editors, wished we had when we were beginning our careers in Washington, DC. Like New York City and Hollywood, Washington is a magnet for idealistic, talented, and ambitious people who have very little idea of what they are actually getting themselves into but nonetheless can't wait to make their mark on the world. The armed forces provide the exception to this rule, as many members of the military actively avoid service in the nation's capital, preferring to serve at "the tip of the spear" than in windowless offices or "cubicle farms" inside the Pentagon. Nevertheless, few military officers can ascend to the ranks of senior leadership without eventually being assigned to Washington, DC.

Regardless of the job or the level of enthusiasm with which they begin, all of these DC newcomers inevitably become emmeshed, flabbergasted, and stymied by entrenched bureaucratic processes they had no idea existed and which often forestall the "obvious solution" to whatever problem they were attempting to solve. This invisible but very real web of processes and authorities constitutes the "rules of the game" for the bureaucracy that underpins American government and governs the actions that can be taken. Thus, a glamorous and exciting career in diplomacy or working on the Hill often involves the process-oriented work necessary to keep departments funded and the government functional. In order to succeed, one must understand these rules, especially as they apply to resourcing. Without funding, strategies and policies are merely interesting ideas. Getting an idea or a program resourced requires a thorough understanding of the process.

Graduate school programs in security studies, public policy, and political science offer multiple courses that consider bureaucracy from academic and theoretical perspectives. That said, these classes generally do not attempt to offer a practitioner's view of surviving and thriving within the Washington bureaucracy.[2] Although individual government departments and agencies such as the Department of Defense's Joint Staff and the Department of State's Foreign Service Institute (FSI) offer courses for personnel newly assigned to Washington, the majority of the learning occurs through on-the-job training.[3]

Considering the national security enterprise from the standpoint of strategic resourcing is neither simple nor straightforward. To succeed requires taking a multidisciplinary approach; working with a team with substantial background knowledge on such diverse and byzantine topics as the Department of Defense's acquisition system, the president's budget submission, and the Federal Emergency Management Agency's National Preparedness Frameworks; and having a basic understanding of macroeconomics. Further, the development of a cohesive and logical narrative is difficult because the Framers' intended checks and balances among the executive and legislative branches effectively preclude the possibility of seamless integration among national security priorities. That said, anyone aspiring to the title of "national security professional" must understand the process in order to be effective. This volume is written with this aspirant in mind.

The prevailing view of strategy formulation among practitioners within the US government today is to define desired ends, select among competing alternatives an approach most likely to lead to those ends, and apply resources that enable that approach to achieve the objective.[4] While some scholars have persuasively argued that strategy development through a balancing of ends, ways, and means is overly simplistic and inevitably leads to bad strategy; the ends, ways, and means model remains the predominant calculus for strategy development within the US national security community.[5] Further, although balancing ends, ways, and means does not guarantee strategic success, a persistent imbalance among them is definitely problematic.

Each chapter in this volume was written by a practitioner with decades of experience working on resourcing issues in Washington. Their perspectives are informed by the cultures of the agencies in which they have worked and the positions they have held. Many currently teach in DC-based graduate degree programs in a variety of disciplines from strategy to economics to organizational leadership. This book is intended as a theoretically grounded yet practical guide for aspiring practitioners who are beginning their careers or seeking careers in the American federal government as well as those who wish to learn more about the inner workings of resourcing the national security enterprise.

CROSSCUTTING THEMES

The chapters in this book reflect how diverse the authors' experiences are and how the bureaucracy itself is not perfectly synchronized, but there are common themes running through the chapters that provide a useful framework to consider. They are:

Understanding Bureaucratic Processes Matters

If one doesn't understand how the bureaucracy actually works, then there is almost no chance that one's particular policy or idea will actually be resourced. In chapter 10, John Ferrari argues that fluency with the systems and processes of the Department of Defense (DoD) and the Office of Management and Budget (OMB) requires thousands of hours of time, even years, to truly master. Although this volume is not a short cut to mastery, it is designed to be a reasonable place to begin the study.

Individuals Matter

There is a joke among Pentagon staff officers that the Pentagon is like an iceberg that is being carried by an army of ants and that everyone who works in the building should understand that they are one of the ants. The implication is that nothing an individual accomplishes will actually change anything, that the momentum created by the ant army is inexorable. The contributors to this volume do not agree with this point of view. Although it is easy to become cynical while working within a large bureaucratic structure, individuals can and regularly do make a difference. In *Essence of Decision*, Graham T. Allison introduces his bureaucratic politics model, which argues that individuals are responsible for every consequential policy decision. Further, Allison argues there are idiosyncratic factors that influence which individual's preferred policy is actually enacted. One of these factors is individual competence within the bureaucratic process.[6] In chapter 4, Jason Galui discusses the tremendous impact of knowledgeable individuals working on the National Security Council (NSC) staff.

The US Economy is Uniquely Strong but Not Inexhaustibly So

In chapter 2, Mark Troutman discusses the history, sources, and limits of the US global economic position. He argues the US position within the global economy is currently challenged by near-peer competitors and that American strategy must be crafted with a clear understanding of the American position within the global economy as well as its limits.

Debt and Deficits Still Matter

Whether or not it involves actual warfighting, competition in the national security space is expensive. Given the resource constraints and priorities of the nation, strategies must be feasible and sustainable over time. Several authors in this book examine the current levels of deficit spending and the increasing levels of mandatory spending as a percentage of the national budget, and they conclude the current trajectory of national spending is unsustainable. Given the United States's position in the global economy, it has the luxury of pushing the limits of deficit spending further than other nations, but the end result will be economic harm if the current pattern is not reversed.

The Executive Branch Dominates the Process, but Congress Matters

Executive branch departments and agencies have outsized influence in the resourcing decisions. This volume devotes four chapters to strategic resourcing in the DoD alone because it accounts for the lion's share of national security spending. Nonetheless, as the founders of the nation intended, Congress's appropriations authority can constrain presidential preferences and actions. Heidi Demarest explains this relationship in detail and makes recommendations to improve the efficiency of the process in chapter 3.

Sometimes Crises are Necessary for Meaningful Change

Many authors in this book discuss the effects of World War II, the Cold War, and 9/11 on the growth and development of the federal government and its bureaucracy. From the National Security Act of 1947 to the Homeland Security Act of 2002, crises have allowed for sweeping bureaucratic and departmental changes that would otherwise have been impossible to implement. Going forward, policymakers must be able to not only successfully navigate crises themselves but also capitalize on the opportunities created for reform.

National Security Guidance Matters Somewhat

Every presidential administration since Ronald Reagan's has published a National Security Strategy (NSS) in compliance with the 1986 Goldwater-Nichols Act.[7] Many executive branch departments and agencies also publish strategy documents explaining how their particular organization will work to fulfill their individual mandates in concert with the NSS. For example, the Pentagon publishes the National Defense Strategy (NDS) and National Military Strategy (NMS) to explicate priorities.[8] National security professionals should read and understand these documents, but they should view these documents as aspirational with goals that are overwhelmingly neither resourced nor prioritized.

The Geopolitical Frame Matters

Whether it's the bipolar nuclear competition during the Cold War, or "Engagement and Enlargement" of democracies during the 1990s, or the "Global War on Terrorism" of the early 2000s, the dominant geopolitical frame will drive resourcing decisions. The current lens is "Great Power Competition" (GPC), which we, the editors, believe will endure for at least the next decade. That said, twenty-first-century GPC does not equal "Cold War 2.0." The United States is still struggling to define the contours of this competition and the major investments necessary to successfully compete globally. Although resourcing traditional military capabilities remains important, the contours of twenty-first-century GPC place greater emphasis on gray-zone competition, cyber capabilities, and critical infrastructure protection than on fighter planes and tanks. Given the current squeeze on discretionary spending and the size of the defense budget, it appears that hard choices among which capabilities to invest in and which to take risks on are unavoidable.

INDIVIDUAL CHAPTER SYNOPSES

Each contributor in this volume writes from their individual perspective and experience resourcing national security. The goal is to provide a guide for navigating the complex and challenging environment of conflicting priorities that constitutes the arena of national security policy making and to make actionable recommendations for change where possible.

Chapter 2 begins with an overview of the relationship between the US economy and national security. Mark Troutman explores the connections between economic prosperity and national security throughout American history, examining the economic factors and political decisions that have allowed the United States to avoid thus far many hard choices between "guns and butter." That said, the combination of increasing mandatory spending and mounting debt will not allow the United States to continue on this path indefinitely. He further posits that the American people, when faced with this choice, will likely be more interested in prosperity than in traditional forms of security, arriving at this conclusion through an exploration of the major components of the federal budget and congressional decision-making. The chapter also considers the economic and security implications of the rise of near-peer competitors and concludes with recommendations for focusing defense spending in light of the rising near-peer competition.

Chapter 3 examines the role of Congress in resourcing American national security from historic and bureaucratic perspectives. Heidi Demarest explains that despite its constitutional role to "provide for the common defense," Congress faces substantial impediments to the effective execution of its legally enshrined mandate. Among these are a substantial information asymmetry with executive branch agencies as well as a static committee structure that effectively precludes a holistic appreciation of national security. She also points to the relative lack of education in the processes of resourcing national security among congressional staffers as significant factors in the successful implementation of legislation. To ameliorate these deficiencies, she recommends a return to biennial budgeting, including changing the default expiration for national security appropriations from one to two years. She also recommends a mandated program of education for congressional staffers to familiarize them with the vocabulary, concepts and processes associated with budgeting for national security. Finally, she advocates for the creation of a select committee on national security affairs to improve synergy between the multiple House and Senate committees.

Chapter 4 is written from the perspective of the executive branch. Jason Galui, a former National Security Council staffer and former member of the President's Council of Economic Advisors, provides not only an insider's view of the process but also recommendations for improvement. It provides historical context for the current national security decision-making process and the evolution of the National Security Council. It also traces the path of a presidential decision from a nascent idea, through the policy process, to the OMB and then into the federal budget, outlining the process involved at each step. This chapter places the greatest emphasis on the "people" part of the process and provides concrete recommendations for aspiring national security professionals to maximize their success navigating the NSC "labyrinth" should they ever find themselves assigned to the organization.

Chapter 5 considers the process by which Congress resources the Department of State (DOS) and how the DOS in turn manages its internal programming and budgeting. Geoffrey Odlum, a veteran diplomat with nearly three decades of experience in the foreign service, posits that although the DOS's resourcing processes allow the department to "muddle through," the muddling comes at a strategic cost to national security. To improve American diplomacy, he provides actionable recommendations, including the creation of a "policy and resource integration unit" within the department to improve overall coherence of policy and resourcing priorities department wide. He also recommends that a comprehensive review and simplification of the authorities designated to subordinate units of the department. Like Heidi Demarest, he also makes recommendations to integrate congressional authorization to achieve a more holistic treatment of national security. Specifically, he advocates for the adoption of "annual, comprehensive national security legislation" that addresses both military and nonmilitary national security accounts together.

Chapter 6 considers the role of United Nations (UN) Peacekeeping Operations in the context of US national security. Although the price of US contributions to UN Peacekeeping is relatively small (~ $ 1–2 billion annually), the return on investment is large.[9] For context, America spends more money on Halloween candy each year than UN peacekeeping.[10] Nonetheless, UN dues assessments have been a consistent hot button issue for Congress. Rebecca Patterson argues that their cost is a "relative bargain," compared to what it would cost the United States to undertake the operations using members of its own armed forces.[11] In a real way, peacekeeping operations offer an opportunity to "outsource" national security a lower cost, both in terms of personnel and dollars.

The Department of Defense, with its highly developed and decades-old Planning, Programming, Budgeting and Execution (PPBE) process, is covered in four chapters in this volume. This is a consequence of the magnitude of DoD spending. The 2021 DoD budget request exceeded $700 billion dollars and represents more than half of the total discretionary budget.[12] Thus, understanding the mechanics of the DoD's resourcing processes is critical to understanding American national security resourcing writ large.

The DoD section begins with a chapter on understanding the defense budget process from the perspectives of the individual services through to the Office of the Secretary of Defense (OSD) to the OMB and then to Congress. In chapter 7, Tom McNaugher emphasizes the fact that the actual budget is an "outcome" of a mammoth process, rather than a desired result. The following chapter by Michael Linick explains the process by which the DoD develops its strategy through the balance of four critical variables: force structure, force posture, readiness, and modernization. He shows how various combinations of these variables in the context of strategic risk constitute the basis for every national defense strategy. In chapter 9, Laura Junor provides us with an in-depth look at military readiness, explaining the complex interplay of macroeconomic factors that influence military readiness as well as the ways that the department evaluates its readiness. The final fourth chapter on the DoD is by John Ferrari, and it is concerned with how the DoD programs its budget and how often, the strategy formulation and programming phases are out of alignment.

In chapter 11, Mark Troutman examines the US government's newest department: the Department of Homeland Security (DHS). Created in the wake of the 9/11 attacks, the DHS is less a single department and more so a loose confederation of departments and agencies operating in conjunction with state and local governments as well as private industries. As a result of both its relative youth and its diffuse and extensive mandates, the process by which DHS creates and resources strategy is simultaneously the least developed and potentially the most consequential of the cabinet agencies. Troutman argues that DHS's mission is likely to grow as a result of the increase in "gray-zone" threats to the homeland but that the department itself is underresourced to adequately fulfill its increasing mission requirements and that the authorities that govern its funding are equally amorphous and result from a tangled web of arcane, often contradictory legislative initiatives.

CONCLUSION

We advise readers to be clear-eyed in their studies of national security resourcing. The process is neither orderly nor integrated. Rather, it is chaotic—careening toward an outcome, pushed by powerful forces and competing agendas. Nonetheless, the authors remain hopeful. Despite its many flaws and in some cases dysfunctional practices, the strategy and resourcing processes of the US government have managed to perform admirably on needed occasions. An immature process mobilized the country to victory in World War II and sustained investment and capability for four decades of the Cold War. After initial misstarts in the twin engagements of Iraq and Afghanistan, the resourcing process provided needed material, human capital, and organizational solutions for the national security community. The experience of crisis response after all is that no capability perfectly matches the arisen need. The aim then is to find those practices that will supply the best strategy resource alignment that will provide the flexibility to adapt as needed and the ability to do it for the long haul.

NOTES

1. Greenberg, *Confessions of A Government Man*, 3.
2. A notable exception to this is *The National Security Enterprise: Navigating the Labyrinth*, edited by Roger George and Harvey Rishikoff. It does an excellent job explaining the major departments and organizational cultures within the American Federal government. That said, it is does not undertake an examination of how national security enterprise is resourced.
3. The Foreign Service Institute offers a four-day long class for Department of State personnel beginning their first assignment in Washington, DC. In the Pentagon, the military services and the Joint Staff offer introductory training for staff officers. However, these courses are short and tend to focus on memoranda formats and broad overviews. For more information, see https://www.state.gov/bureaus-offices/under-secretary-for-management/foreign-service-institute/ and https://www.afms1.belvoir.army.mil/aoic.html for examples and specific curriculum.
4. Arthur F. Lykke, Jr., "Defining Military Strategy," *Military Review* 69, no. 5, (May 1989): 3.
5. Jeffrey W. Meiser, "Are Our Strategic Models Flawed? Ends + Ways + Means = (Bad) Strategy," *Parameters* 46, no. 4, (2016): 82.
6. Allison, *Essence of Decision*, 161–184.
7. The National Security Strategy (NSS) is a report mandated by Section 603 of the Goldwater-Nichols Department of Defense Reorganization Act of 1986 (Public Law 99-433). The NSS has been transmitted annually since 1987.
8. All US National Security Strategies published are available at the US National Security Strategy Archive; see https://nssarchive.us.
9. U.S. Library of Congress, Congressional Research Service, *U.S. Funding to the United Nations System* by Luisa Blanchfield.
10. Center for Strategic and International Studies, "The Price of Peacekeeping."
11. "Peacekeeping: Cost Comparison of Actual UN and Hypothetical Operations in Haiti," U.S. Government Accountability Office, February 2006, https://www.gao.gov/new.items/d06331.pdf.
12. U.S. Library of Congress, Congressional Research Service, *FY 2021 Defense Budget Request* by Brendan W. McGarry.

CHAPTER 2

RESOURCING NATIONAL SECURITY IN A DEMOCRATIC SOCIETY

Mark D. Troutman

> There is no instance of a state having benefited from a prolonged war.
>
> –Sun Tzu, *The Art of War*[1]

INTRODUCTION: CAPACITY AND CONSTRAINT

In 2021 the United States government will spend $740 billion to staff, equip, train, and develop new capabilities for its military. The nation will spend further billions in security investments across other departments that are the subject of this book.[2] The money allocated to the Department of Defense (DoD) is greater than the next ten largest defense budgets worldwide.[3] However, this massive sum represents only slightly over 3% of the US Gross Domestic Product (GDP).[4] Nonetheless, this staggering number should convince readers of the American commitment to its own defense as well as to that of its allies and treaty partners. This approach served the country well from World War II through the Cold War and beyond. But the central question is: Is this approach sustainable over time?

As of the time of this writing, the United States has spent more than seven times its annual defense budget in response to the COVID-19 pandemic, which still has not yet been contained.[5] These two massive expenditures—defense spending and COVID relief—highlight the inherent tension woven into the fabric of US government spending: the demand for individual prosperity versus the requirement for national security. Throughout its history, the United States has relied on its immense natural resource endowment, consistent economic growth, and position of influence in the world economy to realize high living standards while asserting its security priorities. In recent years, this winning combination of factors has eroded and competitors whose potential economic capacity rivals that of the United States have arisen. Long-term trends such as diminishing budget flexibility, slowing growth, and international competition lie at the root of this erosion. Failure to address these trends increases the tension between "guns and butter," and this will force the United States to adjust its priorities in a way that does not comport with the longstanding preferences of its citizens.

Americans want it all—high living standards, physical security, domestic order, and international influence. To achieve this, the United States has long pursued a "rich man's" approach to national security, which leverages US economic capacity to achieve these ends. To effectively practice their craft, strategists and policy makers must understand how the national economy generates the resources that convert into the capabilities (means) to achieve strategic objectives (ends). They must also consider these in the context of future trends and emerging vulnerabilities. This chapter reviews the sources of US economic power, describes how the nation has chosen to harness its economic capacity, assesses current vulnerabilities, and makes recommendations for adjustments required to maintain its current place of global preeminence.

SOURCES OF US ECONOMIC POWER

The United States remains the world's largest economy at $20.9 trillion, even after a 3.5% drop in output from the COVID-19 pandemic.[6] Its residents, who represent a mere 5% of the world's population, enjoy high living standards and a per capita income of over $65,000.[7] US workers generate 15.9% of world output, giving them outsized international influence.[8] Further, over one fifth of the world's largest 500 companies are headquartered in the United States.[9] Despite this substantial wealth, resources are finite. Security must compete among a host of national priorities.

Americans value the economic security embodied in high and improving living standards. James Carville captured this sentiment with his infamous "...it's the economy, stupid..." quip during the 1992 election.[10] The US domestic economy is based on free market capitalism. However, this does not mean that the US government takes a totally "hands-off" stance. The federal government acts through multiple mechanisms to "tune" American economic performance. These include fiscal policy which incorporates taxes, spending, and transfer; monetary policy including the maintenance of bank reserves; and regulation or "rule setting" that defines competitive conditions and standards to compel specific action. These methods, in combination, direct resources to achieve ends that are not delivered in a completely free market.

There are multiple sources of American economic power. Historian John Steele Gordon identifies such elements as: natural resources, a favorable geostrategic position, and an openness to immigration that ensures a young and growing workforce. He also points to a decentralized government structure that created a large unified market, developed flexible yet broad fiscal powers, and made public investments into areas such as infrastructure and education systems.[11] Economist Alan Greenspan and journalist Adrian Wooldridge cite the importance of culture, competitive forces of free markets, investment, and strong institutions such as legal systems and corporate structures that foster innovation and create growth.[12]

All these factors combined to create an American economy that was the largest and fastest growing in the world by the late 1800s.[13] The Bretton Woods economic system, constructed as part of the post–World War II international order, provides the United States with an ability to shape the international financial system on terms favorable to itself and to place the dollar at its epicenter, which has proven to be a significant advantage over the past seven decades.[14]

The US economy has consistently expanded despite periods of temporary contraction, growing at an estimated 6% a year before World War II and at roughly 3% on average during the postwar period.[15] The nation's wealth has provided the country with an ability to "...field an overwhelming force and combine it with economic power and leadership in global affairs to bring to bear far greater resources than any other country against any threat to the nation's security."[16] Economic wealth has also allowed the United States to wield influence in such areas as foreign aid ($40 billion in 2019) and attract substantial levels of international investment through favorable terms of trade.[17] This combination of factors provides the United States the exceptional power to spend from its wealth and borrow when it wishes, all the while exerting immense influence over the international economic system.

THE USES OF ECONOMIC POWER

The United States has consistently invested in public goods that deliver both economic and national security. In 1940, social security issued check 00-000-001 for $22.54. Today, social security delivers to an average retired worker a monthly benefit of over $1,500 at an annual cost of $1.16 trillion or 5.2% of GDP.[18] During roughly the same period, US defense spending increased in real terms from $485 billion in 1953 to $740 billion in 2020.[19] The share of GDP necessary to finance this growing spending has fallen from 9% in 1953 to 3.4% in 2020.[20] The nation's wealth and fiscal policy have thus far furnished both butter and guns. However, the American people have shown an increasing preference for economic security if faced with a choice.

The US government has liberally used its spending, borrowing, and credit power during times of crisis. Federal government spending more than tripled during World War II, and defense budgets grew by 50% during each of the nation's post–World War II conflicts.[21] The Federal Reserve purchased government debt during World War II to ensure low interest rates, expanding credit by fivefold during that conflict.[22] The United States was also able to commit significant federal spending in the closing days of the Cold War, while simultaneously maintaining rising living standards.[23] The nation's ability to marshal economic resources has consistently proven decisive in crisis.

Increasingly, Americans favor the use of borrowing, spending, and credit to preserve jobs and maintain income during economic crisis.[24] The federal bailout associated with the 2008 global financial crisis totaled 8.8% of GDP over five years.[25] The US government response to the COVID-19 pandemic has thus far totaled $5.3 trillion (24.6% of GDP).[26] The Federal Reserve, charged with controlling inflation and supporting employment, boosted its support to the financial system by $3 trillion in response to the Global Financial Crisis.[27] The COVID-19 response has expanded credit by $4 trillion, or 20.1% of GDP.[28] These stimulus packages rival the size of wartime interventions and send a clear message: Americans will hold little back when economic security is threatened.

TRENDS AND THREATS

The United States has relied on its unique advantages and consistently growing economy to provide security and rising living standards. Where Americans have not funded priorities with increasing tax revenues, they have increasingly relied on the nation's borrowing capacity to sustain spending.[29] This has allowed Americans to avoid hard choices, but left unaddressed, this approach will risk diminished future monetary and fiscal flexibility. Moreover, America's growth rate is slowing while China —its nearest peer competitor—maintains higher albeit slowing growth rates. As the United States enters a period of increased economic and great power competition, there is concern about the borrowing capacity and budget flexibility the nation will possess to compete with a rival economy that may surpass the US economy in absolute size within the next decades.[30]

Fiscal and Monetary Flexibility

Americans' high expectations with regard to living standards have set federal government finances on an unsustainable path. US federal spending falls into three broad categories: mandatory spending, interest payments on the national debt, and discretionary spending to fund all other programs. Mandatory spending, which includes social security and healthcare as well as unemployment and poverty assistance is set by law and currently accounts for 61.2% of the federal budget. It increases each year as the baby boomer generation reaches retirement. Interest payments currently require an additional 8.4%.[31] The remaining 30.4% of the budget is comprised of discretionary spending, which is hotly debated annually.[32]

DoD spending encompasses slightly over half of discretionary spending and accounts for 16% of the 2020 federal budget.[33] The remaining federal security-related departments—state, homeland security, and veterans affairs—equate to 15% of discretionary funds and 4.2% of the federal budget. Functions critical to governance and economic growth such as infrastructure investment, education, and research come from the remaining sliver of discretionary spending.[34]

Mandatory spending is growing rapidly and will continue to do so as the US population ages and an increasing number of people qualify for benefits. With interest rates currently at historic lows, debt service costs are also low. Yet as deficits add to the overall debt and interest rates remain on an uncertain upward path, debt service costs run the risk of rapid rise. As a result, the combination of growing mandatory spending and increasing interest payments will further squeeze the shrinking discretionary budget, intensifying the competition for budget funds. Congress may increase revenues or rely on continued borrowing to sustain spending, with recent trends favoring the latter option.[35] Post COVID, the United States is on a path to run persistent deficits exceeding 4% of GDP, with overall debt as high as 225% of GDP in thirty years.[36]

Domestic and international credit markets have so far absorbed large volumes of US Treasury debt without requiring higher interest rates. This introduces the possibility that the United States may both run persistent large deficits and enjoy low interest rates. One view holds that because of the US dollar's status as the dominant reserve currency and demand for the dollar as a "safe haven" asset, activity of the Federal Reserve will continue to absorb treasury debt and hold interest rates low.[37]

Other views hold that low interest rates are an indicator of weak aggregate demand, making deficit spending and resultant debt buildup both low risk and imperative. Sustained economic growth will ease the debt burden, so long as the economy's growth rate exceeds the interest rate financing the debt.[38] However, excessive use of credit brings the risk of higher inflation, which in turn places upward pressure on interest rates.[39] Higher interest rates drive up debt service costs and worsen the "crowding out" effect of mandatory spending and interest costs on discretionary spending.[40]

Further research indicates that the United States may enjoy a range of sustainable debt increase, past which the ability to borrow further begins to deteriorate. One study identifies that debt sustainability begins to deteriorate past 109% of GDP, which implies a long-term sustainable deficit for the US of 2% GDP with maximum debt levels as high as 218% of GDP.[41]

Therefore, while debt service costs remain low, the US fiscal position is sustainable. The federal government has an opportunity to shift spending toward investments that foster growth. Absent economic growth and attendant revenue increase or moderations in spending, the federal budget is on an unsustainable path. Moreover, the nation is entering a period of increased economic and security competition with potentially diminished fiscal and monetary flexibility.

Growth

In the short term (weeks and months), the national economy's potential output is constrained by the size of its workforce and industrial capacity. Over longer timeframes (measured in months to years), the nation can increase its potential output by enlarging its workforce through birth or immigration; increasing industrial capacity by investment in equipment; and improving productivity. In addition to increasing the goods and services available, increased output also dampens inflation.[42] The reality is that US economic growth is slowing, and the trend is continued slowing unless the US addresses the factors that favor growth.[43]

Several factors are known to drive economic growth; these include an enlarging and employed workforce, investment in production capacity, and productivity improvements that increase output from labor and industry. As the US population ages and immigration policy remains contested, the growth of the labor force remains in doubt. Therefore, productivity growth matters even more.[44]

While some factors that drive productivity growth remain the subject of research, investments in education, infrastructure, market efficiency, and institutions that foster competition and innovation are known productivity enhancements.[45] Unfortunately, federal funds for research and development, education, and investment in infrastructure come from the dwindling bucket of discretionary spending.

International Economic Influence

The US remains the world's largest economy by overall output and is among the world's most affluent nations measured by output per capita. The US share of world GDP peaked in 1950 at 28% and has declined to 15.9% as other nations have experienced more rapid growth.[46] China's emergence as a large, technologically capable economy integrated into the world economy and willing to pursue its objectives with tools of economic coercion presents an increasing challenge for the United States.

For example, China's lending to other nations now exceeds the entire loan portfolio of the World Bank.[47] China is also the largest trading nation in the world economy and is a major trade partner of many European and Asian nations who are long-standing American security partners.[48] China's state-led economic policy and use of industrial espionage challenges the rules-based economic system established at Bretton Woods.[49] China, along with Russia, has created financial networks that circumvent the US dollar's role in the international financial system.[50] If sustained, these practices change the US relationship with the world economy in areas of investment flows, trade, fiscal and monetary policy, and business activity. The US ability to sustain economic growth, maintain the vitality of its free-market system, and stand as a compelling alternative to state-directed systems will be vital to the sustainment of its international influence.[51]

THE REAL ECONOMY: THE RESOURCE GENERATOR

Economists commonly refer to the "real economy" of industries that produce goods and services in contrast to the "financial economy" that facilitates savings and investment flows.[52] The real economy is the focus of this chapter, but it is important to note that financial markets are also vital to growth and that financial crises are monumentally expensive.[53] The real economy is the primary source of income as well as goods and services that underpins American economic security. The nation's economy enabled the prosecution of the First and Second World Wars. It also provided an effective counter to the Marxist economic systems of Cold War adversaries.[54] The American private sector also drives innovation, which generates new technologies and production methods. This in turn increases output from labor and capital and accounts for more than half of annual real growth.[55] Further, much of the technology used by the American military originated as private-sector innovation.

US companies continue to lead the world in innovation. Thompson Reuters's 2020 list of leading global technology companies named 45 US firms among the top 100 across financial, legal, operational, social responsibility, and risk measures.[56] The US also scored second behind Singapore in the World Economic Forum's 2019 Global Competitiveness Index, which assesses the institutions, policies, and factors that determine national productivity.[57] It is significant that US firms in the Thompson Reuters's top 100 list also appear on a list of promising artificial intelligence (AI) companies, a technology of increasing significance in terms of national security.[58] Innovative industry does triple duty by developing industries that deliver economic security; creating key technology related to national security, and generating a network of influence that projects US power.

Several elements support the development of competitive and highly productive US firms. Among these are intellectual property protection and investment funding. Intellectual property results from research and development and translates into new products and improved methods that directly contribute to economic growth. Increasingly, intellectual property (IP) is digital and relies on cyber security for protection.[59] Today, IP theft is rampant, lucrative, and often conducted by foreign government intelligence services. It is estimated to cost US businesses $600 billion per year.[60] IP theft is an illegal wealth transfer that decreases return on investment, threatens economic growth, and diminishes the ability to maintain technological and productivity advantages.

US industry is a constant target of attack, both to extract value and to cause physical damage. The 2015 attack on the Sony Corporation, attributed to North Korean operatives, is an early case of digital theft.[61] More recent attempts to steal information on US defense programs and the COVID-19 vaccine have direct national security impacts.[62] Industry is vulnerable to digital intrusion which can take many different forms, from attacks on power grids to attacks on the US water supply.[63] A lucrative and attractive target for both private competitors and foreign adversaries, US private industry is literally under direct attack.

Robust investment by the public and private sector sustains both innovation and the productivity of US industry. US research and development (R&D) expenditures accounted for $476.5 billion (2.7% of GDP) in 2020. These investments support a research ecosystem comprising companies, research centers, and academic institutions.[64] Economists Johnathan Gruber and Simon Johnson identify a "crowding-in" effect whereby public research funds draw private funds into industry and result in job growth and breakthrough research. One estimate quantifies the effect— each 1% increase in defense research funds results in an increase of 0.5% in private-industry research funding.[65] A powerful example of this effect is the Human Genome Project, which the federal government funded in its early research stages. As commercial applications became more apparent, private-sector companies have increasingly provided investment funds.[66] Yet federal research funds have declined from a high of 2.2% of GDP in 1964 to 0.6% currently.[67] Federal R&D funding has also suffered from the competition within discretionary spending. In 1965, federal R&D funding was 25% of discretionary spending; in 2020 it was only 11%.[68]

THE DEFENSE INDUSTRY

US defense industry relies on an ever-smaller number of companies as technology has become more specialized.[69] Some observers such as strategists Cleveland, Bryant, David, and Jensen point out that the defense industry has the tendency to focus on platforms oriented toward high-intensity warfare, which represents expensive investments in low-probability, high-consequence options.[70] As defense technology has become more specialized, concerns have grown within the US government regarding the growth of program costs, defense supply chain security, and the ability of the US industrial base to surge in case of emergency.[71] A second set of concerns surrounds the defense sector's ability to incorporate the broader economy's innovation and productivity into its massive commitment of resources.

The procurement and R&D portions of the 2020 DoD budget total $244 billion, and the department issued $421.5 billion in contracts to private industry. While huge, these sums are small when compared to spending in other sectors such as the automobile industry, which boasts annual revenues of $953 billion.[72] As a result, some companies face the dilemma of participating in the defense sector, with its specialized, often frustrating procedures, and priorities that conflict at times with private-sector opportunities.[73]

The modern defense industry has become increasingly privatized and concentrated in the hands of a few mammoth companies.[74] Following the end of the Cold War and collapse of the Soviet Union, the US reduced defense spending and the defense industry consolidated, going from 58 companies to five major defense-focused "primes" with a reduced supply chain of sub-tier firms. This left a more concentrated and specialized defense industry without the depth and diversity of its Cold War predecessor.[75] Concerns arise over a lack of competition with and resulting deficiencies in performance, delivery, and cost control. In 2020, the General Accounting Office (GAO) reported that major defense acquisition programs cost grew at rates greater than inflation and required delivery times of 30% longer than initial program guidelines.[76] The industry also faces problems of supplier depth as well as labor force challenges. It also has reduced expansion potential, particularly in those systems with a limited commercial counterpart.[77]

Multiple initiatives have sought to reduce cost, improve system performance, and shrink delays, among them the Better Buying Power (BBP) initiatives, rapid contracting authorities in the 2018 National Defense Authorization Act and the Adaptive Acquisition Framework.[78] Rapid procurement systems such as the Agile Acquisition Framework (AAF), research structures such as Defense Advanced Research Projects Agency (DARPA) and funding vehicles such as In-Q-Tel incorporate private-industry vehicles such as venture capital to integrate cutting-edge technologies into the DoD.[79] The expanded authorities are in greater use over the last decade and have drawn more and new companies into the defense network. Much remains to be done as the majority of contract dollars still flows to traditional defense companies.[80]

In spite of it all, the defense acquisition system and its industry partners do field systems that are in demand. Last year, US defense exports totaled $175.1 billion, which was the largest of any nation. This number illustrates the enduring competitiveness of US systems on world markets.[81] Despite its imperfections, the US defense industry and its acquisition system may represent a "least bad" solution given the complexity of both the product and the system involved. Nonetheless, a clear case for reform that delivers defense capabilities more quickly and at less expense exists.

CONCLUSION

Strategists and policy makers must consider multiple elements of economic capacity, performance, industry competitiveness, and future trends to effectively practice their craft. There are clearly vulnerabilities that the US must address if it is to continue to raise living standards and remain competitive on the world stage. Economic capacity resources national security. Clearly, competitor nations seek to challenge US economic performance and displace the leadership the nation has known for the past seven decades.

US citizens have become accustomed to increasing living standards, robust responses to crisis, and increasing investment in national defense. Recent threats to US industry and the global pandemic have struck at economic capacity. Currently, the nation faces slowing growth that threatens to derail the capabilities that have allowed it to exert influence. Economic growth is fundamental to satisfy these simultaneous demands and requires investments in infrastructure, immigration, education, research funding, and intellectual property protection. A stable and growing economy that operates on favorable terms in the international system allows for the use of fiscal and monetary tools to support strategic objectives. The nation has a near-term opportunity with low debt costs to make investments that foster growth, which will allow it to place fiscal and monetary policies on sustainable paths in the medium and long term. The US can leverage these investments by cultivating relationships with allies and trade partners to pool R&D funds, counterbalance competitor-nation pressures, and offer market-based alternatives to authoritarian economic approaches.

Challenged in the international arena, US industry faces widespread IP theft and is under attack in the digital realm. The US should protect industry by seeking international agreements that protect IP and favor competition over managed market solutions. The nation should invest in measures that protect digital traffic and cultivate international relationships that enforce common protection standards. In particular, the US should protect its defense industry from digital theft; the recently deployed Cybersecurity Maturity Model Certification (CMMC) standards are a positive step. The US should negotiate international agreement that forbids digital attacks on critical infrastructure industries and use its cyber capabilities to expose and impose costs on actors who attack commercial activity.

The United States should increase investment in future capabilities such as high-speed broadband communications and curtail programs that are not relevant to future missions. Responsive acquisition and mobilization structures must draw the best technology and companies into the national security realm. Growth, investment, and a revitalization of market forces that foster productivity improvement are the surest way to maintain the vibrancy of US economic capacity. The most significant investment into national security will prove to be a secure, innovative, growing, and resilient national economy.

NOTES

1. Sun Tzu, *The Art of War*, chapter 2.
2. Office of Management and Budget, "Budget of the U.S. Government, Fiscal Year 2020."
3. Peterson, "U.S. Defense Spending Compared to Other Countries.".
4. Calculations based on table B2 of Council of Economic Advisors, *Economic Report of the President* (2020); and Peterson, "U.S. Defense Spending Compared to Other Countries."
5. Peterson, "Here's Everything the Federal Government Has Done to Respond to the Coronavirus So Far." Secretary of the Treasury Stephen Mnuchin remarked to Chris Roberts of Fox News on the first stimulus package "...we're at war..." on Fox News Sunday, April 26, 2021, https:// heisenbergreport.com/2020/04/26/steve-mnuchin-on-deficit-national-debt-were-at-war-chris/comment-page-1/#:~:text=Steve%20Mnuchin%2 0On%20Deficit%2C%20National%20Debt%3A%20%E2%80%98We%E2%8 0%99re%20At,long%20claimed%20is%20a%20road%20to%20financial%20 oblivion.
6. Bureau of Economic Analysis, "Gross Domestic Product, Fourth Quarter and Year 2020."
7. World Bank, "GDP per Capita (Current US$)—United States." https:// data.worldbank.org/indicator/NY.GDP.PCAP.CD?locations=US.
8. Ibid.
9. Rapp, "Visualizing the Global Fortune 500 (2020)."
10. Zelizer, "'It's the Economy, Stupid' All over Again.".
11. Gordon, *An Empire of Wealth*, xiii–xviii.
12. Greenspan and Wooldridge, *Capitalism in America*, 419.
13. Maddison, *The World Economy*, 262.
14. Steil, *The Battle of Bretton Woods*, chapter 8.
15. Smits, Woltjer, and Ma, "A Dataset on Comparative Historical National Accounts, 1870–1950"; and Council of Economic Advisors, *Economic Report of the President* (2020), table B-1.
16. U.S. Library of Congress, Congressional Research Service, *Economics and National Security*, by Richard K. Nanto, 4–5.
17. Ingram, "What Every American Should Know About US Foreign Aid."
18. Author calculations for FY 2020 based on Congressional Budget Office, *The Budget and Economic Outlook: 2021 to 2031*," 2–4; "Social

Security History," https://www.ssa.gov/history/index.html and "Monthly Statistical Snapshot," Social Security Administration, June 2021. The term entitlement refers to any payment due to an individual who qualifies by requirements outlined in law. See also Cogan, *The High Cost of Good Intentions*, 79–81.

19. Office of the Under Secretary of Defense (Comptroller), *National Defense Budget Estimates (2021)*, 80.
20. Council of Economic Advisers, *Economic Report of the President* (2021), table B-47.
21. Ibid.; and Office of the Under Secretary of Defense (Comptroller). *National Defense Budget Estimates for FY 2020 (Green Book)*, 80.
22. Waller and Ricketts, "Fed Balance Sheet."
23. Bowie and Immerman, *Waging Peace*, 96–108.
24. The Savings and Loan Crisis of the late 1980s and early 1990s involved costs of $153B or a stimulus of 1.5% of GDP; the early 2000 recession required a $100 billion stimulus (1.0% GDP). Shibut and Curry, "Cost of the Savings and Loan Crisis," 34; and CNN Money, "Bush Seeks Up to 75B."
25. Kessler, "Did Wall Street Get a 'Trillion-Dollar Bailout' during the Financial Crisis?."
26. Peterson, "Here's Everything the Federal Government Has Done to Respond to the Coronavirus So Far."
27. Stamborski, "A Look at the Fed's Dual Mandate"; and Waller and Ricketts, "Fed Balance Sheet".
28. Board of Governors of the Federal Reserve System, "Credit and Liquidity Programs and the Balance Sheet."
29. The 2020 deficit is $3.1 trillion; it will decline in the early part of this decade and then increase with projected spending and revenue patterns. See Swagel, "The 2021 Long-Term Budget Outlook."
30. Zhu and Orlik, "When Will China Rule the World? Maybe Never.".
31. The US government could suspend or reschedule interest payments, with a likely downgrade of its credit rating as happened in July 2011, and the potential for higher interest rates. British Broadcasting Corporation, "US Loses AAA Credit Rating after S&P Downgrade."
32. Swagel, "The 2021 Long-Term Budget Outlook," 18.
33. *Economic Report of the President* (2020), table B-47.
34. Ibid., and Swagel, "The 2021 Long-Term Budget Outlook," 18.
35. Ibid.

36. Auerbach and Gale, "The Effects of the COVID Pandemic on the Federal Budget Outlook."
37. See Fullweiler, "The Debt Ratio and Sustainable Economic Policy."
38. Furman and Summers, A Reconsideration of Fiscal Policy in the Era of Low Interest Rates."
39. Friedman and Schwartz, *A Monetary History of the United States*, 676–678.
40. The consensus view favors low inflation, though there is a growing group of economists who warn that the combination of large stimulus and credit expansion at the same time risks taking recovery past full economic capacity and thereby igniting inflation. See Summers, "The Biden stimulus is admirably ambitious. But it brings some big risks, too."
41. Mian, Straub, and Sufi, "A Goldilocks Theory of Fiscal Policy."
42. See chapters 8 and 9 in Mankiw, *Macroeconomics*.
43. Swagel, "The 2021 Long-Term Budget Outlook."
44. Ibid.
45. Kim and Loayza, "Productivity Growth: Patterns and Determinants across the World"; and Henderson, "Economic Growth."
46. Bhutada, "The U.S. Share of the Global Economy Over Time."
47. "The Overseas Activities of Chinese Banks Shift Up a Gear," *The Economist.*
48. Wolf, "Containing China Is Not a Feasible Option."
49. "Xi Jinping Is Trying to Remake the Chinese Economy," *The Economist.*
50. Greenwald, *The Future of the United States Dollar.*
51. Blackwill and Harris, *War by Other Means*, 220 – 226.
52. Oxford English Dictionary, "The Real Economy," https://dictionary.cambridge.org/dictionary/english/real-economy
53. See note 26 of this chapter on financial interventions. See also Kiel and Nguyen, "Bailout Tracker."
54. Gordon, *An Empire of Wealth*, 413–415.
55. Henderson, "Economic Growth."
56. Paladino, "Top 100 Global Tech Leaders."
57. Schwab and Zahidi, "Global Competitiveness Report 2020."
58. "10 Artificial Intelligence Companies Leading the Smart Revolution," *Analytics Insight News*
59. Manyika et al., "Digital Globalization: The New Era of Global Flows."
60. Lewis, "The Economic Impact of Cybercrime."
61. Buchanan, *The Hacker and the State*, 167–186.
62. Gould, "Wide Reaching Hack Has Defense Firms on Their Toes."

63. Cybersecurity and Infrastructure Security Agency, "Cyber-Attack Against Ukrainian Critical Infrastructure"; and MacColl and Dawda, "US Water Plant Suffers Cyber Attack Through the Front Door."
64. UNESCO Institute for Statistics, "How Much Does Your Country Invest in R&D?."
65. Moretti, Steinwender, and Van Reenen. "The Intellectual Spoils of War?,", 26, 35.
66. Gruber and Johnson, *Jump-Starting America*, 113–136.
67. Atkinsonand Foote. "Federal Support for R&D Continues Its Ignominious Slide."
68. American Association for the Advancement of Science, "Historical Trends in Federal R&D."
69. Cancian and Saxton, "Industrial Mobilization," 21–22.
70. Cleveland et al., *Military Strategy for the 21st Century*, 63–89.
71. Interagency Task Force in Fulfillment of Executive Order 13806, "Assessing and Strengthening the Manufacturing and Defense Industrial Base and Supply Chain Resiliency of the United States."
72. "America's Automobile Industry Is One of the Most Powerful Engines Driving the US Economy," Alliance of Automobile Manufacturers.
73. Hsu, "Pentagon Warns Silicon Valley About Aiding Chinese Military."
74. Interagency Task Force in Fulfillment of Executive Order 13806, "Assessing and Strengthening the Manufacturing and Defense Industrial Base and Supply Chain Resiliency of the United States."
75. Watts, "The US Defense Industrial Base," 2.
76. U.S. Government Accountability Office, "Defense Acquisitions Annual Assessment: Drive to Deliver Capabilities Faster Increases Importance of Program Knowledge and Consistent Data for Oversight." Hereafter "Defense Acquisitions Annual Assessment 2020."
77. Interagency Task Force in Fulfillment of Executive Order 13806, "Assessing and Strengthening the Manufacturing and Defense Industrial Base and Supply Chain Resiliency of the United States," 48–49.
78. U.S. Library of Congress, Congressional Research Service, *Acquisition Reform in the FY2016-FY2018 National Defense Authorization Acts (NDAAs)*, by Heidi M. Peters.
79. U.S. Library of Congress, Congressional Research Service, *Department of Defense Use of Other Transaction Authority: Background, Analysis, and Issues for Congress*, by Heidi M Peters.
80. Ibid.

81. U.S. Department of Defense, "DOS and DOD Officials Brief Reporters on Fiscal 2020 Arms Transfer Figures.

CHAPTER 3

THE ROLE OF CONGRESS

Heidi Demarest

But each proposal must be weighed in the light of a broader
consideration: the need to maintain balance in and among national
programs—balance between the private and public economy,
balance between cost and hoped for advantage—balance between
the clearly necessary and the comfortably desirable; balance
between our essential requirements as a nation and the duties
imposed by the nation upon the individual; balance between
actions of a moment and the national welfare of the future.
—Dwight D. Eisenhower,
Farewell Address, January 17, 1961[1]

INTRODUCTION

The Founders wisely created a government underpinned by a system
of checks and balances. As a result, executive agencies are powerless
to resource programs to implement a strategy absent a congressional

appropriation signed into law. This chapter explores the history of federal budgeting, or the backdrop against which all decisions about allocating taxpayer dollars to national priorities are made. Despite Congress's constitutional responsibility to provide for the common defense, significant impediments exist to exerting legislative voice and influence over the direction of national security. Among them, Congress faces an inevitable asymmetry of information with executive agencies, particularly within the complex, opaque Department of Defense. Congress has also maintained over the past century a relatively static committee structure that fragments a holistic perspective on national security budgeting and inhibits the ability to identify, evaluate, and deliberate trade space among programs. Finally, constituents' particularized interests over the short term, coupled with an annual budget process, disincentivize many Congress members from considering and providing oversight of long-term strategic choices proposed by federal agencies—for who among the American public, absent, perhaps, national security professionals, can determine whether their member of Congress has made the right decision about buying fighter jets versus nuclear submarines?

Potential solutions to remedy the legislative branch's handicaps in the arena of resourcing national security include a robust training and education program for relevant members of Congress and their personal staff on national security threats, strategy, and existing programs; creating a select committee on national security affairs to evaluate long-term consequences and tradeoffs of budgetary decisions pertaining to national security; a return to biennial budgeting to promote better fiscal discipline and enable longer-term strategic thinking among decision makers; and, ultimately, education of the electorate to demand a more sustainable view of budgeting for defense and the ability to hold elected officials accountable for their constitutionally mandated role. [2] Budgeteers and programmers in executive agencies provide critical expertise to turn resources into capability, but the ultimate architects of national security are the voting public who elect the chosen few to carry out the business of aligning ends and means to strategy on the public's behalf.

FEDERAL BUDGETING HAS BECOME MORE FRAGMENTED AND LESS DISCIPLINED

How Congress budgets for national security is related to how Congress approaches federal budgeting, writ large. Denizens of the modern era, inured to the terms "deficit spending," "national debt," and "sequestration," may not realize that our forebears were far more accustomed to the concepts of "balanced budgets," "public credit," and "regular redemption and discharge of the public debt."[3] Alexander Hamilton was a staunch advocate of building national credit, creating federal revenue streams and avoiding excessive interest on money borrowed from other nations. In *Federalist*, no. 12, Hamilton declares the self-evident truth that

A nation cannot long exist without revenues. Destitute of this essential support, it must resign

its independence and sink into the degraded condition of a province. This is an extremity to which no government will of choice accede. Revenue, therefore, must be had at all events.[4]

The United States even paid off its entire debt, once, under President Andrew Jackson and the 24th Congress in 1835. In the moment, Jackson's achievement represented the American ideals of freedom and exceptionalism. No other country had accomplished such a feat.[5] And although it was a "memorable and happy event," the moment was brief.[6]

President Jackson is given most of the credit for eliminating the national debt in 1835, but Congress shared the Framers' disinclination toward profligate spending. They designed the federal budgeting process with the principles of relative balance between revenues and expenditures at the forefront during the early years of the Republic. Congress dominated federal budgeting in the nineteenth century. In those years, all financial matters, including all revenue and spending bills, were concentrated in the House Ways and Means and the Senate Finance Committees.[7] But in the late 1800s, after unprecedented expansion of the American state during the Civil War, Congress decided to assign responsibility for revenue bills and spending bills to different committees. They federated the spending bills even further by tasking general topics to various legislative committees, with the result that the House and Senate appropriations committees no longer maintained a holistic view of the nation's expenditures.[8] Federal agencies submitted their budgets directly to these various congressional committees, using their own organizing principles and interests as the basis of their budget submissions.

The balance of power in federal budgeting began to shift away from Congress toward the president with the Budget Act of 1921 that followed the major expansion of the federal government with the prosecution of World War I and the permanent establishment of an income tax. Congress reassigned all responsibility for spending bills back to the appropriations committees but required the president to submit an annual consolidated budget to Congress. The Budget Act provided the president with a staff, today known as the Office of Management and Budget, to help establish a process for consolidating and reviewing agency budget submissions.[9] Since 1921, presidents have set initial annual budgetary priorities and recommended allocations to programs, including national security programs, for congressional review.

When the government spends more money than it collects, the difference is called a deficit. Countries borrow money from the public or other nations to make up their deficits and then pay interest on the borrowed amounts. Unsurprisingly, war drives up spending and deficits. World War II was the most expensive conflict in US history, increasing American debt for decades. More permanent deficits in America unconnected to war costs began to balloon in the 1960s, primarily due to expanding mandatory spending bills for statutory entitlements like social security and Medicare. Absent new legislation, Congress *must* pay these bills each year, which account for approximately 60 percent of the nation's annual expenditures. About 35 percent of the budget is discretionary, over which the president and Congress maintain far more control. National defense spending typically represents half of all discretionary spending.[10] Unsurprisingly, war drives up spending and deficits. World War II was the most expensive conflict in United States history, followed by the 2003 Iraq War and the Vietnam War.[11] Defense spending usually declines precipitously after major armed conflicts, but generally not to prewar levels. In fiscal year 2020, the US Congress appropriated $738 billion to national defense, which amounts to a contribution of about $2,400 per citizen.

Today, Congress attempts to impose discipline on the budget process by using budget resolutions to set funding levels by functions of government, though these do not have the force of law. The Budget Control Act of 2011 set spending caps on discretionary spending and the public debt for the subsequent decade, but Congress has habitually either raised the ceilings or suspended them, negating the control measures.[12] The evolution of federal budgeting provides useful context when considering the defense budget, or any activities related to congressional appropriations for national security purposes. Over the past 200 years, the share of discretionary spending as a proportion of the federal budget has declined dramatically, federal spending has grown by orders of magnitude, and the power to define the budget's contours has shifted away from Congress toward the agencies of the executive branch. Congress faces further challenges when deciding how to allocate the American citizen's $2,400 annual contribution to national defense.

INFORMATION ASYMMETRY

An enduring problem across congressional budget functions is the amount of time members of Congress and their staff can devote to assessing the president's proposed budget when it arrives on the Hill on the first Monday of February.[13] Members of Congress are busy, and each year they confront an imperative to pass a budget by the time the prior year's appropriations expire on September 30. Congress members are aided by the professional staff members of the House and Senate armed services committees, which authorize funds for national defense, and the House and Senate Appropriations Committees (subcommittees on defense), which make appropriations for national defense. Professional staff members have often served for many years studying the same portfolio and are more conversant about historical trends, schedules, costs, and performance issues of national security programs than Pentagon officials.[14] However, the Department of Defense submits a budget that routinely exceeds 1,700 pages and takes several hundred people over nine months to prepare. Given this, the twelve to fourteen professional committee staffers on the appropriations subcommittees on defense and approximately forty staffers on the armed services committees are simply at a natural disadvantage in terms of numbers and time to detect anomalies, errors in calculations, or substantive objections to the proposed array of resources.

The complexity of information and Congress members' relative lack of familiarity with defense programs impairs oversight and analysis of alternatives to the Department of Defense's annual budget submission. As Frank Baumgartner and Bryan Jones put it, "The more you look for problems, the more you find. The more you seek to understand the complexity of a given problem, the more complex you find that it is."[15] A career bureaucrat in the Department of the Army who has spent the last thirty years costing and analyzing the engine that powers the Blackhawk helicopter is likely intimately familiar with the program's technical history, contracting structure, and manufacturing concerns. A congressional staffer, particularly on a Congress member's personal staff, is simply unable to devote the time to acquiring and analyzing the information required to discern potential issues with a given budget line item, much less synthesizing that information across multiple programs. This is not a new problem for congressional oversight. Historically, Congress relies on citizens and organizations to set off so-called "fire alarms" when federal agencies violate congressional intent.[16] This approach has the benefit of conserving limited congressional time and attention, but it does not enable congressional staff to weave the president's initial budget proposal into a cohesive account to address the nation's national security requirements or to adjust the account substantially.

The problem of information asymmetry between Congress and the executive branch is exacerbated by trends in congressional staffing over the past several decades. The pay for Congress members' personal staffers has declined since the 1990s, shifting away from policy or legislative experts in favor of relatively cheaper labor oriented on constituent services.[17] A 2020 study on congressional staff pay and tenure finds that "most staff that manage policy portfolios in Congress have only one or two years of experience. That is, roughly one-third of legislative staffers have not yet served the duration of a single Congress."[18] Certainly, professional committee staff members and Congress members' senior staff directors are typically more experienced and more highly compensated, with higher levels of expertise in their assigned portfolios of legislative responsibility than the typical staffer or intern in a Congress member's personal office. But Congress members' access to information and analysis to support their constitutional responsibilities in the realm of providing for the common defense is variable at best. The consequences of asymmetrical information are even more pronounced in the face of Congress's structural challenges to achieve a comprehensive view of the national security picture.

SUBOPTIMAL ORGANIZATIONAL STRUCTURE

The budget process is a major driver of collective congressional activity and dictates the calendar of authorizing and appropriating committees. Each year the House and Senate budget committees report a concurrent budget resolution, which establishes spending limits for the various functions of government, currently broken into twenty-one categories. As discussed earlier, the budget resolution does not have the force of law. Function 050 encompasses activities dedicated to national defense, including "the military activities of the Department of Defense (DoD), the nuclear-weapons-related activities of the Department of Energy (DOE) and the National Nuclear Security Administration (NNSA), the national security activities of several other agencies such as the Selective Service Agency and portions of the activities of the Coast Guard and the

Federal Bureau of Investigation."[19] Budget function 150, International Affairs, includes programming for military and economic foreign aid as well as commitments to international organizations and peacekeeping operations. The budget resolution is one place where "resourcing for national security" is considered in the aggregate against other competing priorities for funding. Once the budget resolution is adopted, it is further divided into allocations for each of the twelve subcommittees within the House and Senate appropriations committees. Each subcommittee eventually reports out an appropriations bill to the full committee or may combine their bill with that of other subcommittees to form an omnibus spending bill.

Federal agencies cannot allocate money to programs without a congressional appropriation. The president's proposed budget, which is submitted to Congress each February, is examined by both authorizing committees and appropriations committees responsible for oversight of agency operations, spending and functions. The House and Senate armed services committees have jurisdiction over national defense and typically orient their review of the budget on the policy implications of the president's spending choices. The House and Senate appropriations subcommittees on defense examine the budget in parallel with the armed services committees. Should the versions of the authorization and appropriations bills differ between chambers, Congress members will conduct a conference to reconcile the bills, put them to a floor vote, and eventually present the president with a National Defense Authorization Act and a National Defense Appropriations Act to sign into law.

A chief advantage of this process is its orderly and predictable subdivisions of jurisdiction, which partially assists committees and their staff to counter the information asymmetry problem identified earlier. Professional staff members for the armed services and appropriations subcommittees on defense are typically longstanding public servants with previous expertise in the national security enterprise.[20] There is less turnover among professional staff members, and many of them retain institutional memory extending beyond that of Department of Defense officials and military officers (career defense bureaucrats notwithstanding). And the process does assign ultimate control for all spending bills to the full House and Senate appropriations committees, an improvement over the fractured and unwieldy situation prior to the Budget and Accounting Act of 1921.[21]

Centralized congressional budgeting with decentralized responsibility along functional lines is efficient but acts as a double-edged sword. Committee deliberations about proposed US military exercises in the Pacific may occur in the Senate armed services subcommittee on readiness, while the proposed distribution of foreign aid to Indo-Pacific nations occurs in the Senate foreign relations committee and trade policy with China and Taiwan is debated in the Senate finance committee. Further, the budget resolution and its functional spending targets are the product of the House and Senate budget committees. These committees are charged not only with delineating the contours of federal spending but reconciling spending with revenues and enforcing budget rules that have imposed discipline on the budget process over the years. Budget committee staff members have likened their committee to "the skunk at the garden party."[22] The budget resolution is a product of prevailing national economic conditions, anticipated expenditures versus receipts, and the estimated fiscal impact of any legislation likely to be considered by congressional committees.[23] It is not the product of a holistic debate over national strategy or a document that reflects all the components of national power contributing to a strategic direction for national security, measured deliberately against the nation's competing priorities and commitments. The closest we have to such a product is the annual Appropriations Act signed into law, which in the aggregate tends to resemble the president's initial budget proposal closely.

CONGRESSIONAL BUDGETING: SHORT-TERM GAINS OVER LONG-TERM INVESTMENTS

Legislators' fundamental goal in seeking reelection is another obstacle to the optimal alignment of means to ends in national security. Political scientists have long characterized members of Congress as single-minded seekers of reelection—regardless of other goals that legislators may have, reelection is "the proximate goal of everyone, the goal that must be achieved over and over if other ends are to be entertained."[24] Congress members are unable to affect the budget process or the strategic direction of national security policy if they are not in office. The election cycle dominates Congress members' time and attention, particularly in the House, where terms are only a third of the length of a senator's term. The election cycle influences Congress members' budgeting decisions, the amount of attention they devote to matters of national security and their perception of what is important to their constituents; for example, Congress members are more likely to approve generous appropriations in an election year and more likely to make program cuts following an election year.[25] Providing oversight of fiscal discipline within the Department of Defense, reviewing the US Army's wheeled-vehicle replacement strategy and discussing the Joint Staff's realignment of forces available to combatant commanders may or may not be at the forefront of their task list as the calendar approaches November.

Time is also a factor in agencies' ability to expend appropriated funds. Congress can specify the time for which an appropriation will remain available to an agency. The money must be obligated for expenditure prior to its expiration date. Some appropriations are designated as "no-year," meaning they are not tied to obligation in any particular fiscal year, while others are "multi-year," or exist for use in more than one fiscal year but expire on a designated date. Unless otherwise specified in law, all appropriations are annual. Over half of the federal discretionary budget consists of annual appropriations, and about 65 percent of the Department of Defense appropriations are one-year money.[26] There are many advantages to annual budget authority—chief among them is that it promotes legislators' ability to affect the direction of the budget from year to year, consistent with their oversight responsibilities. But annual budgeting also incentivizes agencies to spend soon-to-expire appropriations at the end of the fiscal year, rather than return unobligated funds to the US Treasury. If agencies do not spend their entire appropriated budget, they face the very real possibility of a smaller allocation in the following fiscal year. An analysis of the timing in federal purchases from fiscal years 2004 to 2009 shows that end-of-year spending on nonessential purchases such as information technology, furniture, and building maintenance occurs at a rate nearly five times the spending rate during the rest of the year.[27] Many agencies are required to submit their obligation plans for their entire budget allocation at the outset of the fiscal year and have instituted procedures to monitor the execution of funds throughout the year, but the practice of end-of-year spending is persistent and widespread.[28] Over the long term, one-year appropriations arguably erode Congress's ability to influence national security agencies to make resourcing decisions consistent with long-term strategic investments.

Finally, the strategic choices in the National Defense Authorization Act and National Defense Appropriations Act are not well understood by most Congress members' constituents, hampering the public's ability to hold lawmakers accountable for their budgetary decisions. Congress members understand this phenomenon and act accordingly. Debates over base closures and military construction attract relatively more Congress member involvement than most weapons procurement decisions. While Congress members are more likely to defer to defense officials' expertise on the technical requirements of missiles and helicopters intended to defeat an adversary, there is little doubt of their active attention to the Department of Defense's selection process for the location of a new military command in Texas or Massachusetts.[29]

This is not to say that average citizens do not possess opinions about defense spending or are unable to communicate their opinions when asked for them. Citizens' preferences for defense spending do seem to take primacy over preferences for domestic programs, meaning that most people express a stronger preference for either more or less defense spending than a preference for more or less domestic spending. As you may expect, opinions about defense spending are affected by the salience of an external threat, such as the former Soviet Union, but citizens have typically supported increased spending levels for both defense and social programs over time.[30] Congress members may gauge their constituents' attitudes toward aggregate defense spending or collect polling data on the use of military force, but these attitudes and preferences are not narrow enough for Congress members to discern their constituents' reaction to their potential support of the army's personnel management strategy or the number of ships in the navy's fleet. If the costs and benefits of a policy cannot be perceived by the public, attentiveness declines.[31] What matters most to constituents matters most to Congress members, and most constituents simply do not know about the particulars of their annual $2,400 contribution to national defense. Undeniably, some Congress members are personally interested in national security policy and the application of resources required to support it. Hearings, requests for information from defense agencies, and reports or inquiries into federal agencies' stewardship of national security programs are all useful tools to elicit valuable information at a relatively low cost to Congress members and their staff. But in general, it is not a close tether of electoral accountability that incentivizes Congress members to develop deep expertise in national security programs.

RECOMMENDATIONS

Information asymmetry exists between the executive and legislative branches, with the balance of deep technical and institutional knowledge falling to the bureaucracy. The congressional committee structure and budgeting process fracture Congress's ability to make deliberate trade-offs and cohesive strategic decisions about resourcing national security.

Electoral incentives, the time horizon of annual appropriations, and an inattentive public combine to create circumstances favoring decisions with short-term payoffs rather than long-term investments. Given these significant challenges, what solutions could better enable the legislative branch's constitutional responsibility to provide for the common defense?

First, Congress should create a training program for Congress members and their staff to accelerate familiarization with recurring concepts and vocabulary in the national security arena. Professional development and certification programs for congressional staffers is not a new idea.[32] An onboarding process exists for first-term Congress members and their staff, along with myriad resources for conducting legislative operations, performing constituent services and familiarization with professional protocols and norms.[33] Committees have discretion to conduct their own local versions of onboarding for new professional staff members, who possess relative longevity compared to the tenure of personal staff members. But high turnover coupled with the complexity of national security programs necessitates a more intensive, targeted investment in members of Congress and their legislative assistants, particularly Congress members who are on authorizing and appropriating committees for national security functions. At a minimum, curriculum should include information about how to interpret budget-justification materials, an overview of agency responsibilities for specific national security programs, familiarization with key national security documents, an accompanying critique from scholars in the security studies and political science disciplines, an assessment of the current and future national security environment from the intelligence community, and an informal discussion with current military and administration officials responsible for enacting national security policy.

Second, the congressional committee structure as it pertains to national security should be reorganized to address the currently fragmented view of authorizing and appropriating committees. A select committee on National Security Affairs could provide "oversight and responsibility of the joint space between Congress and the executive branch" and improve the totality of the legislative branch's view while allocating resources to national security.[34] This solution could present a valuable synergy between House and Senate foreign relations committees, budget committees, armed services committees, intelligence committees, and appropriations committees. The select committee would require the authority to affect decisions about resourcing if it is to act as a credible principal in the constellation of agency oversight, and contain enough resident expertise to deliberate in a substantive and productive way about aligning national means to strategic ends.[35] The executive branch invested in a major reorganization in the wake of 9/11 to address gaps in responsibility between agencies, particularly those with a mission to analyze and disseminate intelligence. The legislative branch has not had a major reorganization since the Legislative Reorganization Act of 1970.[36] A thoughtful effort to move Congress members into seeing beyond the end of their current term and promote debate within the budgetary trade space can help strengthen congressional oversight and result in better informed national security programming.

Third, Congress should reexamine the benefits of a biennial budgeting cycle, including changing the default expiration date for national security appropriations from one to two years. Doubling the time horizon for use of funds, particularly those unrelated to pay and allowances, will reduce the potential for profligate spending against unnecessary priorities at the end of every fiscal year. Two-year appropriations may improve the ability of both the legislative and executive branches to make longer-term investments and more strategic choices. Biennial budgeting has long been discussed by Congress and the executive branch and is currently employed by nineteen states.[37] Critics of the two-year cycle argue that fact-of-life changes that occur during the off-years will be so numerous as to negate any efficiencies gained from extending the annual appropriations cycle and that eliminating the annual budget discussion between Congress and the executive branch further weakens the legislative oversight function. Proponents are more optimistic about the benefits to both planning and fiscal management within Congress and executive agencies by relieving both branches of the time-intensive requirements to propose, defend, examine, and pass an annual budget. An extended time horizon permits Congress to conduct "more meaningful program reviews" and become more familiar with the complexities and interrelated tradeoffs within the budget.[38] Further, aligning the budget cycle with the election cycle could establish more direct and responsive electoral accountability for budgetary decision-making in the national security arena.

CONCLUSION

The US Congress holds the constitutional responsibility for resourcing national security. Congress faces numerous obstacles to exercising meaningful oversight of executive agencies and influencing national security programs, including information deficits, a fragmented committee system that stovepipes decisions better made in a holistic strategic fashion, and electoral incentives that do not necessarily align with the longer-term strategic goals of national security policy. Several solutions to these challenges are presented here, including additional training and professional certifications for Congress members and their staff in the realm of national security programs, an overhaul of the congressional committee structure to eliminate gaps between multiple committees responsible for portions of the nation's comprehensive approach to national security, and the institution of a biennial budget cycle to enable longer-term predictability and blunt the annual short-term incentive structure in both the legislative and executive branches. This is by no means an exhaustive list, and alternatives proliferate among scholars and practitioners alike. But the most challenging obstacle to resourcing national security is, perhaps, an inattentive public. In the absence of informed constituents who send a signal to their elected representatives that they *do* care about tradeoffs in providing for the common defense, Congress members are left to make the inevitable conclusion that their time is better spent on other policy areas and will simply do the best they can. As President

Eisenhower warned, "only an alert and knowledgeable citizenry" has the power to compel republican governments to act in alignment with the nation's goals. As in most areas of republican government, responsibility begins with the citizens, who may find themselves more inclined to learn about the tradeoffs between fighter jets and nuclear submarines.

NOTES

1. Eisenhower, "Farewell Address," January 17, 1961.
2. Demarest and Borghard, eds., Afterword to *US National Security Reform*, 205–211.
3. From George Washington's perspective, "I entertain a strong hope that the state of the National Finances is now sufficiently matured to enable you to enter upon systematic and effectual arrangements for the regular redemption and discharge of the public debt, according to the right, which has been reserved to the Government. No measure can be regarded as more desirable whether viewed with an eye to its intrinsic importance or to the general sentiment and wish of the Nation." Draft of George Washington's Fourth Annual Address to Congress, October 15–31, 1792, National Archives: Founders Online, accessed September 1, 2020, https://founders.archives.gov/documents/Hamilton/01-12-02-0393#ARHN-01-12-02-0393-fn-0020.
4. Hamilton, *Federalist*, no. 12, 91.
5. Lane, "The Elimination of the National Debt in 1835," 76.
6. "I cannot too cordially congratulate Congress and my fellow citizens on the near approach of that memorable and happy event—the extinction of the public debt of this great and free nation." Andrew Jackson, "Fourth Annual Message to Congress," December 4, 1832, https://millercenter.org/the-presidency/presidential-speeches/december-4-1832-fourth-annual-message-congress.
7. Schick, *The Federal Budget*, 34.
8. Ibid., 35.
9. Congressional Budget Office, "History."
10. Congressional Budget Office, "Discretionary Spending Options."
11. World War II cost the United States approximately $4.1 trillion dollars; the Vietnam War $738 billion, and the Iraq War approximately $1.6 trillion dollars as of 2014. Daggett, *Costs of Major U.S. Wars*, 2. See also Belasco, *The Cost of Iraq, Afghanistan, and Other Global War on Terror Operations Since 9/11*.
12. The Budget Act of 2019 suspended the debt ceiling limit for FY20 and FY21, effectively nullifying the final two years of the Budget Control Act of 2011. U.S. Congress, H.R. 3877, "Bipartisan Budget Act of 2019," https://www.congress.gov/bill/116th-congress/house-bill/3877#:~:text=%2F02

%2F2019),Bipartisan%20Budget%20Act%20of%202019,for%20defense%20 and%20nondefense%20spending.

13. The president's budget is often not delivered to Congress on time, further compressing the congressional authorization and appropriation process. Office of the Management and Budget, *Circular No. A-11: Preparation, Submission, and Execution of the Budget,* December 2019, Section 10, 4.

14. Demarest, *US Defense Budget Outcomes,* 56–59.

15. Baumgartner and Jones, *The Politics of Information,* 3.

16. McCubbins and Schwartz, "Congressional Oversight Overlooked," 172.

17. Crosson et.al., "Partisan Competition and the Decline in Legislative Capacity among Congressional Offices."

18. Furnas and LaPira, "Congressional Brain Drain: Legislative Capacity in the 21st Century," 25.

19. "Budget Functions," House Committee on the Budget, accessed September 12, 2020, https://budget.house.gov/budgets/budget-functions.

20. Demarest, *US Defense Budget Outcomes,* 57–58.

21. U.S. Senate Budget Committee, "Committee on the Budget, United States Senate, 1974–2006," 109th Congress, 2nd session, 2006, S. Doc. 109-24, 22.

22. Senate Budget Committee, "Committee on the Budget," 104.

23. Heniff, *Formulation and Content of the Budget Resolution,* 1.

24. Mayhew, *Congress: The Electoral Connection,* 16.

25. Kiewiet and McCubbins, "Congressional Appropriations and the Electoral Connection," 79. See also Demarest, *US Defense Budget Outcomes,* 129–131.

26. Congressional Budget Office, Letter to Representative Womack, Re: Period of Availability of Appropriated Funds, May 21, 2018, https://www.cbo.gov/system/files/2018-07/54155-appropriationsletter.pdf.

27. Liebman and Mahoney, "Do Expiring Budgets Lead to Wasteful Year-End Spending? Evidence from Federal Procurement," 2.

28. Many federal watchdog groups and federal contracting interest groups routinely track this phenomenon, including FedSmith.com, GovSpend.com, and the National Taxpayers Union. All unclassified federal spending data is available at https://www.usaspending.gov/..

29. Goss, "Military Committee Membership and Defense-Related Benefits in the House of Representatives," 232.

30. Wlezien, "The Public as Thermostat: Dynamics of Preferences for Spending," 997.

31. R. Douglas Arnold, *The Logic of Congressional Action,* 28.

32. Furnas and LaPira, "Congressional Brain Drain," 46–47. See also Demarest and Borghard, eds., *US National Security Reform*, 206–207.

33. For instance, see the Congressional Management Foundation series https://www.congressfoundation.org/congressional-operations/new-member-resource-center/new-member-training-series and the Senate New Member Orientation, https://www.senate.gov/about/traditions-symbols/orientation-programs.htm.

34. Demarest and Borghard, eds., *US National Security Reform*, 207.

35. Heidi Kitrosser, "Congressional Oversight of National Security Activities," 107.

36. The Legislative Reorganization Act of 1970 established the Congressional Research Service and increased staffing in the U.S. Government Accountability Office, along with professionalizing committee staff. Kravitz, "The Advent of the Modern Congress," 383.

37. Saturno, *Biennial Budgeting: Issues, Options, and Congressional*, 13. As this report mentions, the state experience with biennial budgeting has not provided enough data to reach a clear answer about whether it is superior to annual budgeting. See also U.S. Government Accountability Office, *Biennial Budgeting: Three States' Experiences.*

38. Saturno, *Biennial Budgeting*, 6.

CHAPTER 4

RESOURCING STRATEGY AND POLICY

THE NATIONAL SECURITY COUNCIL

Jason Galui

> For the Romans, as for ourselves, the elusive goal of strategic statecraft was to provide security for the civilization without prejudicing the vitality of its economic base and without compromising the stability of an evolving political order.
>
> — Edward Luttwak,
> *The Grand Strategy of the Roman Empire*[1]

INTRODUCTION

Edward Luttwak's description of the elusive goal of strategic statecraft succinctly articulates the challenge facing every government; that is, a government must allocate its scarce resources across tools of national power to provide for the common defense and promote the general welfare. For the United States, does a more-than-$700 billion defense budget provide sufficient security while preserving the vitality of its economic base or compromising a stable evolving political order? Would reallocation of federal funding from defense priorities to education reform, for instance, provide enough security while enhancing the "national welfare of the future"?[2] There are no "right solutions" to complex national security challenges, especially when juxtaposed with complex domestic policy issues.

This chapter provides a perspective from the central node of US decision-making, the National Security Council (NSC), to explore how the United States develops policy choices at the highest level. It will identify processes and people as the key obstacles and constraints to achieving better strategic clarity (or less strategic confusion) and offer practical recommendations for small teams and individuals to improve their contributions to the US national security decision-making process.

When presenting options to the president, US senior leaders must present actions that increase the likelihood of achieving more preferred outcomes along with the resources available to pursuing those actions. Simply stated, strategy and budget resourcing must align. Any strategy or policy without available resources to execute becomes primarily an interesting academic exercise that frustrates the goal of effective presidential policymaking. This chapter is devoted to exploring how processes and people can introduce strategic confusion into US national security and foreign policy. It concludes with some practical recommendations to teams and individuals on how they can reduce strategic confusion within their component of the national security apparatus. It will first explore how the White House develops and coordinates strategy and policy across departments and agencies. Then, it briefly delves into the federal budget process from a White House perspective overlaying this process to national security decision-making.

THE NATIONAL SECURITY COUNCIL PROCESS

Since the passage of the National Security Act of 1947, US national security decision-making has become an increasingly systematized affair. As national security challenges grow increasingly complex, the number of stakeholders and influencers for any issue rises, while available resources to address those national security challenges grow increasingly scarce. Moreover, the boundary between foreign policy and domestic policy is eroding; that is, assuming it ever existed.[3]

While every president has adapted White House decision-making processes to fit their unique decision-making styles and personal tendencies, the formal NSC structure and process remains relatively unchanged since President George H.W. Bush occupied the Oval Office and Lieutenant General Brent Scowcroft led the NSC in the late 1980s and early 1990s. Scowcroft viewed the role of the National Security Advisor principally as that of a custodian of a policy process rather than a policy advocate. Scowcroft formally established the Principals Committee (PC) and Deputies Committee (DC) to support the NSC, and in turn Policy Coordination Committees (PCC) supported meetings of the PC and DC.[4] The strength of any particular policy depends on how well the Policy Coordination Committees/ Interagency Policy Committees (PCC/IPC) achieves an effective and efficient coordination process.

Formal membership of the PC has varied slightly across administrations, though the chair of the PC has tended to be the National Security Advisor. Cabinet secretaries and directors from relevant agencies, including the Chairman of the Joint Chiefs of Staff, are members of the PC and are collectively known as the "principals." Key White House officials such as the Assistant to the President and White House Chief of Staff; the directors of the Office of Management and Budget (OMB), the National Economic Council (NEC), and the Office of Science and Technology Policy; and the chairman of the Council of Economic Advisers may attend PC meetings depending on the relevancy of the topic.[5] Deputy secretaries and deputy directors from relevant agencies along with the vice chairman of the Joint

Chiefs of Staff comprise the DC, and collectively are referred to as the "deputies." Similar to the PC, appropriate White House component deputy directors may attend DC meetings at the discretion of the meeting chair.

Before an issue reaches principals or deputies, the issue iterates between sub-PCCs/IPCs and PCCs/IPCs (Policy Coordinating Committee/Interagency Policy Committee). Substantive experts and senior officials—at the levels of assistant secretary and two- or three-star flag officers—attend the PCC/IPC, which is chaired by an NSC senior director. Participants in the PCC/IPC carry a large responsibility into their meetings because they often are the lone representative (perhaps with one additional teammate in support) for an agency's analysis and proposed recommendations on a particular issue. These interagency partners must be well-informed on their agency's capabilities and capacity to bring to bear on a national security challenge. At the PCC/IPC, these participants represent the ways and means to be allocated toward a particular policy,

Policy synchronization and management of policy implementation occurs within the PCC/IPC, and within the supporting sub-PCC/sub-IPCs. These are generally chaired by an NSC director. It is at this level that any "whole-of-government approach" is formulated. A "successful" interagency working group at the IPC/PCC, led by a strong NSC senior director serving as an honest broker, can build a strong bridge linking strategy, policy, and resources. Failure at the IPC/PCC level results in a weak "strategy bridge" and thus contributes to increased strategic confusion as result of disconnections between policy and resources.

ALLOCATING RESOURCES

Any discussion about strategy and policy formulation divorced from a corresponding discussion on allocation of resources to implement strategy and policy is incomplete. The US budget process was discussed in detail in chapter 3. However, a very brief overview of that process focusing on the critical role of the Office of Management and Budget

(OMB) is essential for understanding the resourcing of national security from the perspective of National Security Council.

The US Constitution provides the Congress the power "to lay and collect taxes, ... to pay the debts and provide for the common Defence [*sic*] and general Welfare of the United States; ...." The Constitution, however, does not prescribe how such legislative processes are to be exercised, nor does it provide a specific role for the president regarding budgetary matters. Various statutes, congressional rules, practices, and precedents have been established over time to create a complex system in which multiple decisions and actions occur with varying degrees of coordination.[6] Perhaps most notable among them is the Budget and Accounting Act of 1921, which created a statutory role for the president by requiring executive departments and agencies to submit their budget requests to the president and, in turn, for the president to submit a consolidated request to the Congress. The simplest description of today's US budget cycle consists of four complex stages: 1) the President's Budget formulation and submission to the Congress (after the first Monday in January, but not later than the first Monday in February); 2) congressional consideration of budgetary measures; 3) budget execution; and 4) audit and review. OMB plays a central role in the formulation of the President's Budget, the management of budget execution by departments and agencies, and interagency coordination of policy initiatives.[7]

If the NSC is the strategy bridge connecting ends and ways, then the OMB represents the structural support of the NSC strategy bridge. The OMB is the largest component within the Executive Office of the President and serves as a clearinghouse for federal agencies on budget issues and congressional testimony ensuring consistency across the administration. With the federal fiscal year running from October 1 to September 30, the budget process is well-defined and driven by time. A new budget process begins almost immediately upon the submission of the President's Budget to Congress each February.

The president, the OMB director, and other senior administration officials gather in the Oval Office not long after the President's Budget submission for the upcoming year to discuss the president's budget priorities for the fiscal year after next. This late winter or early spring meeting with the OMB director and other senior administration officials begins the roughly 18-month process toward the next fiscal year federal budget. At this meeting, the president issues guidance on matters such as policy priorities as well as acceptable levels of deficit (or surplus) spending and corresponding national debt increases (or reductions). The OMB then delivers letters to departments and agencies, which provide presidential budget guidance for the next budget request. A common point of guidance is to issue a budget cap for a department of agency. For instance, following the 1997 agreement between the president and the Congress to balance the federal budget by 2002, President Bill Clinton issued a $270.8 billion budget ceiling for the Department of Defense in fiscal year 1999.[8]

After notification of the presidential parameters, departments and agencies assemble their budget requests through the summer and fall subject to the fiscal guidance from the OMB and given their respective agency priorities. When complete, departments and agencies submit their budget requests to OMB examiners who review the requests to ensure compliance with the president's guidance. The OMB then returns updated budget targets to the departments and agencies during the passback period. It is during the passback that department and agencies essentially negotiate with the OMB over any differences between an agency's request and the OMB target. Cabinet secretaries and agency directors may even directly contact the OMB director, the White House Chief of Staff, or even the president to discuss their budgetary concerns. Finally, the President's Budget request is submitted to the Congress, which then begins the congressional budget processes for the upcoming fiscal year.[9]

While the federal budget process tracks along a well-defined timeline, the NSC staff manages several policy processes simultaneously, few of which track a similarly well-defined timeline. The NSC processes, however, cannot neglect the federal budget cycle lest they risk pursuit of infeasible options.

DEFINING A SUCCESSFUL NSC DECISION-MAKING PROCESS

A successful national security decision-making process is one that deliberatively, inclusively, and justly generates a full range of feasible options for the president. Achieving such success is at least a two-step process. Step one encompasses policy formulation; that is, from problem identification through policy selection. Step two encompasses policy implementation, evaluation, and revision. Successful implementation and evaluation of policy is itself an entire field of political science and successful implementation of a policy often relies as much, or more, on exogenous factors such as the role of external uncertainty and the "interplay of possibilities, probabilities, good luck and bad"[10] as it does on ones within the control of the policy makers.

National security decision-making, like any other political process, is an inherently messy endeavor consisting of a myriad of actors with disparate policy goals wielding varying levels of influence over the process in the hope of achieving parochial "success" for their specific agency. Although a successful policy-making process in no way guarantees ultimate policy outcome success, it absolutely improves the odds of such success. A successful policy process achieves two interdependent goals: functional and relational. The functional aspect is one that generates a wide array of feasible policy options for the president. The relational aspect is a process that is deliberative, inclusive, and just.

Functional Success

The president is the ultimate decision maker when it comes to protecting the American people and advancing US national interests. The role of the NSC staff is to coordinate with and enable cooperation among departments and agencies to generate the policy options from which the president can select. The success of the NSC is therefore linked to its functional ability to present the president with the maximum number of feasible policy options, ranging across the full spectrum of US resources and response capabilities. To achieve such success, the policy-making process must effectively incorporate the breadth of expertise within US departments and agencies as well as expertise outside of government as appropriate.

A strong PCC/IPC synthesizes the available information and expertise in a systematic and collaborative way, develops feasible policy options to address a national security challenge, and articulates the options succinctly to deputies and principals. "Feasible" refers to options that are within the known constraints of department and agency capabilities and capacities stemming in large part from budget allocation and time available. Thus, the PCC/IPC must be a strong strategy bridge with well-built supports, including OMB participation, that consistently and systematically links potential policy options with available resources.

Relational Success

Strategy making is inherently a human endeavor, and so any strong PCC/IPC must comprise strong working relationships among its members. It is not enough that the policy-making process can produce an adequate range of feasible options for dealing with one policy problem. A successful process must reliably do so in response to every problem the NSC addresses. Thus, the relational aspect of success is just as essential as the functional one. Therefore, the success of the NSC is also contingent on a decision-making process that is perceived by all participants as deliberative, inclusive, and just.

A "deliberative" process is one that establishes a clear definition of the problem, promotes shared understanding of that definition, and then generates potential policy responses in a systematic way. An "inclusive" process is one that solicits input from all relevant departments and agencies that have the expertise and equities that contribute to the generation of potential policy options. A "just" process is one that produces a policy decision that all participants understand and accept, even if some participants disagree with the final decisions. A just process does not necessarily produce consensus, but it does produce a satisfaction among the key participants in the process that their position was fairly and appropriately considered.

If the policy process breaks down, strategic confusion begins to emerge; this happens whenever the process fails to pass the functional or relational tests. Deliberative and inclusive policy processes are more likely than non-deliberative and exclusive ones to generate a full range of feasible options. Thus, participants in interagency meetings must be team players recognizing the role they play for their agency, and consequently the national tool of power that they represent. National security decision-making is the ultimate team sport, and each national security challenge deserves a team of experts committed to a deliberative, inclusive, and just process that generates a full range of feasible options for the president.

THE NSC PROCESS AROUND THE POST-2016 US TROOP PRESENCE IN AFGHANISTAN

A concrete example of the relatively abstract processes and concepts discussed thus far can be found in President Barack Obama's announcement in May 2014 about the US post-2016 troop presence in Afghanistan—the military force component of a complete plan of action for Afghanistan. In his remarks, the president reaffirmed US objectives, which were "disrupting the threats posed by al Qaeda; supporting Afghan security forces; and giving the Afghan people the opportunity to succeed as they stand on their own."[11] President Obama then described the ways and means toward achieving those objectives. The president elaborated that at the start of 2015 the United States would have fewer than 10,000 troops in Afghanistan and that by the end of 2016 the US military would draw down to "a normal embassy presence in Kabul." He clearly articulated the military resources necessary for this updated policy.

Within the NSC staff, this May 2014 announcement simultaneously marked the culmination of one policy process and the start of another. Having achieved policy formulation on future US military presence, the NSC director responsible for US security policy in Afghanistan now moved to overseeing policy implementation and measuring the trajectory of policy outcomes. Complicating the political situation in Afghanistan was the 2014 Afghan presidential election that resulted in the creation of the National Unity Government in which Ashraf Ghani became president and his chief political rival, Abdullah Abdullah, was named Chief Executive Officer. It was a power-sharing agreement facilitated by the United States as a way to maintain peace between Ghani and Abdullah who had competing visions for the future of Afghanistan.

By late 2014 and into early 2015, interagency participants of the Afghanistan sub-IPC began reporting that some of the underlying assumptions upon which the May 2014 announcement of the post-2016 decision had been made were proving invalid. There was broad agreement that the security situation in Afghanistan was deteriorating, and interagency colleagues acknowledged at sub-IPCs and at the IPC that the current post-2016 plan for Afghanistan was unlikely to achieve US objectives. Interagency colleagues pushed on the NSC staff to reexamine the post-2016 US presence. America's longest war was not on a path to a responsible end with the current strategy and policy.[12]

The NSC Afghanistan team recognized both the constraints on the policy-making process as well as the strength of the policy process. While they would raise the concerns with senior leadership of the NSC, they encouraged interagency colleagues to raise similar concerns with agency principals and deputies who, in turn, could raise the concerns with their NSC counterparts, the National Security Advisor and Deputy National Security Advisor. This was an example of the relational aspect of national security decision-making in action. The NSC director had heard interagency colleagues' concerns through inclusive and deliberative dialogue—formally and informally—and was able to reexamine US policy options objectively.

Meanwhile, the OMB examiner responsible for the Afghanistan portfolio was aware of the ongoing process to reexamine US policy options on post-2016 Afghanistan. Maintaining close contact with NSC director as well as interagency counterparts was essential to ensuring any recommended changes to US policy in Afghanistan remained within already appropriated budget levels and within expected future presidential budget requests. Moreover, the NSC director realized the advantage of maintaining close contact with the OMB examiner so that the NSC director could maintain focus on the development of US post-2016 Afghanistan strategy and policy. The OMB examiner personified the NSC strategy bridge structural support to this particular NSC policy process.

As departments and agencies managed their internal processes to raise concerns with their respective principals and deputies, the sub-IPC and IPC process moved forward with development of options for the president's consideration that would alter the drawdown of US forces in Afghanistan. A strong working relationship between the NSC director and the OMB examiner—specifically through multiple daily meetings and phone calls—increased the likelihood of achieving a successful NSC policy process and thus resulted in a broad range of feasible options to the president for his decision.

In March 2015, President Obama amended his May 2014 announcement, noting that US forces in Afghanistan would remain at 9,800 for all of 2015, and in October 2015 he announced his decision to maintain 9,800 US troops in Afghanistan for most of 2016 and to draw down to 5,500 troops "at a small number of bases."[13] As an inclusive, deliberative, and just NSC policy process on Afghanistan continued, the president made his final

Afghanistan announcement in July 2016, noting that US troops would drop only to 8,400 by the end of 2016.[14] These presidential decisions had been well-informed by an NSC policy process that was well-supported by the OMB's role as connective tissue to department and agency resource availability. The success of the interagency process leading to President Obama's final policy decision on post-2016 US presence in Afghanistan was the direct result of the strong relationships between the people involved in the process.

This example of Afghanistan has been discussed not to make a judgement on the policy outcome, but rather to highlight the importance of people and process as constraints on the national security decision-making process. National-level decisions result from the combination of people, process, and policy. To improve strategic competence, we must begin at the individual level by building stronger relationships with colleagues working on similar issues and by understanding the various processes through which we participate; both those internal to one's respective department or agency and those pertaining to how one's department or agency fits into the broader national security decision-making process.

CONCLUSION

The ability of the United States to achieve its national objectives depends not only on US actions but also on the actions of allies, partners, and adversaries. Similarly, ensuring a successful national security decision-making process depends not only on the actions of any one person, department, or agency but on the relationships and contributions of everyone involved in the process. Understanding that everyone working in the executive branch exists, in some form or fashion, to enable a full range of feasible options to be presented to the president of the United States for decision-making is a good starting place to increasing the likelihood of achieving greater strategic competence.

There are so many competing priorities but such limited resources that solving the national constrained optimization problem is nearly impossible. In a democratic system of government, with so many influential actors, public policy-making and policy execution are messy. The more national security professionals understand the national security decision-making process, how their particular agency processes interact within the larger process, and what their individual roles are within the process, the more strategic clarity—or less strategic confusion—will emerge.

While an effective NSC staff is necessary to have a well-built strategy bridge for any policy process, such effective national security professionals are not sufficient to ensure a strong bridge endures. There are many other people involved in the strategy-making process, and all of them must pull their weight to build, support, and maintain strong strategy bridges. Every individual, team, and organization within the national security enterprise has a responsibility to improve the strategic competence of the United States by building stronger "strategy bridges" between policy options and the resources available to realize more preferred outcomes.

As individuals, national security professionals must know their portfolio, for they may be the only person in a department of agency thinking constantly about a particular issue. For instance, the NSC director for Afghanistan security was the only person at the White House always thinking about that portfolio because the NSC senior director had to think about all of South Asia while the National Security Advisor's portfolio almost mirrors the president's portfolio of all matters.

Individuals and teams must also develop relationships within and beyond their respective departments and agencies. Given that people, process, and policy combine to generate national decisions, it is essential to know the people thinking about the same issues and portfolios from other perspectives, especially from other departments and agencies. National security decision-making is the ultimate team sport, and thus national security professionals should seek to build teams across offices and agencies.

Teams and individuals must also know the organization of their department or agency. Understanding bureaucratic processes and how decisions are made within agencies as well as how information flows in, around, and out of respective organizations will enable more effective, efficient, and influential advice and recommendations.

With virtually no "closed-form solutions" to complex national security challenges, allocating scarce resources across tools of national power to advance national interests is a challenging endeavor for any group of people and processes. While the White House and the Executive Office of the President is the central node of strategy development and the builder of strategy bridges linking policy choices to available resources, the federal departments and agencies are charged to recommend feasible options and execute strategy and policy. The executive branch certainly has the leading role in determining preferred allocation of resources in the government's national-level constrained-optimization problem. The president may be the ultimate decider at the end of the foreign policy and national security decision-making processes, but the 535-member US "board of directors" that is the Congress may be the ultimate constraint on the execution and implementation of any presidential policy choice.

NOTES

1. Luttwak, *The Grand Strategy of the Roman Empire,* 94.
2. Eisenhower, Dwight D. *Farewell Address,* January 17, 1961, https://www. eisenhowerlibrary.gov/research/online-documents/farewell-address.
3. President Joe Biden's selection of Ambassador Susan Rice as Director of the Domestic Policy Council provides some support to the notion of erosion of any previously existing boundary between foreign and domestic policy.
4. The Obama Administration referred to Policy Coordination Committees as Interagency Policy Committees (IPCs).
5. The Biden administration published "National Security Memorandum 2" on Feb 4, 2021. It lays out the current membership and invitees to the various level National Security Council Meetings. See https://fas. org/irp/offdocs/nsm/index.html
6. For a detailed description on the development and operation of the framework for budgetary decision-making that occurs today, see Saturno, "Introduction to the Federal Budget Process."
7. Ibid., 14. This process is also described in detail in chapters 3 and 6 of this volume.
8. Wilson, *This War Really Matters,* 84.
9. For a description of the budgetary roles of Congress, see chapter 3 in this volume.
10. Clausewitz, *On War,* 86.
11. White House Office of the Press Secretary, "Statement by the President on Afghanistan," May 27, 2014, https://obamawhitehouse.archives.gov/ the-press-office/2014/05/27/statement-president-afghanistan.
12. The policy process is recounted from my firsthand observations while serving on the staff of the National Security Council.
13. White House Office of the Press Secretary, "Statement by the President on Afghanistan," October 15, 2016, https://obamawhitehouse.archives. gov/the-press-office/2015/10/15/statement-president-afghanistan.
14. Thomas, "Afghanistan: Background and Policy," 8.

CHAPTER 5

RESOURCING US
DIPLOMATIC PRIORITIES

Geoffrey Odlum

If you don't fund the State Department fully, then I need to buy more ammunition ultimately.

—General James Mattis,
Congressional Testimony, 2013[1]

INTRODUCTION

Diplomacy and foreign assistance may be considered the overarching tool of state power: The lead or coordinating instrument within which economic, sanctions, military force, negotiations on specific issues, or any of the other instruments may be placed.[2] When compared to the US government's military and intelligence budgets, diplomacy and foreign assistance are relatively low-cost tools for promoting American values globally; building and maintaining alliances; assessing potential threats from abroad; and providing political, economic, humanitarian, and security assistance to partner countries to support shared interests. However, as this chapter shows, the development of diplomatic and foreign assistance strategy at the Department of State and the US Agency for International Development (USAID) is largely disconnected from the resource planning and budgeting process. Absent a more concerted integration of policy planning and resourcing—a more strategic coordination of diplomatic "ends, ways, and means"—US diplomacy and foreign aid run the risk of becoming ends unto themselves, divorced from grand strategy, diverting scarce resources for ineffective purposes and in the worst case potentially undermining US values and interests abroad.

This chapter describes the foreign policy and foreign assistance planning process and the separate processes for requesting and appropriating resources. It then identifies five key obstacles to integrating policy and resource planning within Congress, in the Executive branch "interagency" process, and within the State Department. This chapter offers seven recommendations to overcome those obstacles.

HISTORY

The conduct of American diplomacy predates the United States, tracing its history to the inception of the Continental Congress. Many of the Founding Fathers, including Benjamin Franklin, Thomas Jefferson, and John Adams, served abroad during the American Revolutionary War as diplomats representing the revolutionary cause, seeking foreign support for America's struggle for independence from British rule. Their experiences convinced them that effective diplomacy was a vital tool to safeguard the fledgling Republic's survival. It was the Founders' clear intent, as affirmed in the US Constitution, that the president would determine US foreign policy.

This intent was reaffirmed in subsequent Acts of Congress which established the Department of State in 1789, making it the oldest executive department of the US government and providing that the Secretary of State would serve as the president's principal foreign policy advisor, with foreign policy carried out by the Department of State and the Diplomatic Service of the United States.[3] It was a function and department so important that James Madison proposed it to the First Congress and Thomas Jefferson was appointed as the first Secretary of State, overseeing a staff of one chief clerk, three junior clerks, a translator and a messenger while directing the work of two diplomatic posts in London and Paris and ten consular posts.[4]

Since its founding, the State Department's core responsibilities have not changed significantly, though they became far more complex as the United States assumed a leadership role in global affairs after World War II, as the number of nation-states and international organizations in the world multiplied and as threats to US interests proliferated. The duties and responsibilities of the Secretary of State and the Department of State include:

- Serving as the President's principal adviser on US foreign policy;
- Conducting negotiations relating to US foreign affairs;
- Advising the President on the appointment of US ambassadors, ministers, consuls, and other diplomatic representatives;
- Participating in international conferences, organizations, and agencies;
- Negotiating, implementing, and terminating treaties and agreements;
- Ensuring US Government protection to American citizens, property, and interests in foreign countries;
- Providing information to American citizens regarding the political, economic, social, cultural, and humanitarian conditions in foreign countries;

- Informing the Congress and American citizens on the conduct of US foreign relations;
- Promoting beneficial economic intercourse between the United States and other countries.[5]

For over 150 years, the Department of State was the principal agency handling American foreign policy, foreign affairs, diplomacy, and foreign assistance. The National Security Act of 1947, however, introduced the broader supplanting concept of "national security" and birthed a far more complex "national security enterprise," bringing together the US government's diplomatic, defense, intelligence, law enforcement, and economic policy agencies under the coordinating umbrella of the National Security Council and its staff.

This had profound impacts on the development of US foreign policy, not least of which was to disperse some traditional foreign policy missions, including aspects of foreign assistance and international economic relationships like finance and trade, to other agencies in the interagency community.[6] Moreover, the new NSC-led process forced the State Department to compete directly with more influential sister agencies (especially the Defense Department and Central Intelligence Agency) for presidential attention, NSC support and congressional interest in diplomatic policies, programs and resources.[7]

DESIGNING FOREIGN POLICY

Modern US foreign policy is formulated in three overlapping ways: (1) An incoming presidential administration typically produces a National Security Strategy (NSS) near the beginning of its time in office. Such strategies look ahead over the administration's term in office; these are written in broad, declaratory, and aspirational terms based on enduring US interests and American values but offer little guidance on how to weigh policy tradeoffs or assess resource requirements. The National Security Strategy serves as the starting point for lower-level diplomatic and foreign assistance plans developed by the State Department and subsidiary units (bureaus and missions) which provide more specific implementation goals. (2) In response to real-world events, the State Department participates actively in the interagency policy-formulation process, the mechanics of which differ from administration to administration but are usually spelled out in a Presidential Decision Directive (PDD), Presidential Policy Directive (PPD), or National Security Policy Memorandum (NSPM). The Biden administration outlined its interagency process in National Security Memorandum-2 (NSM-2) issued on February 4, 2021.[8] (3) Impromptu policy decisions made by the president in real time, sometimes announced directly to the public by means of social media.

This approach to foreign policy decision-making adds unpredictability to the policy-planning and resourcing process. Sudden shifts in US foreign policy are fully within the authority of the president to decide, but they often preclude an opportunity for effective review by the interagency or careful assessment of resource implications and tradeoffs.

THE STATE DEPARTMENT'S FOREIGN POLICY PLANNING PROCESS

All national security policy–planning processes begin with strategic guidance from the National Security Strategy. The NSS describes US national security interests, objectives, and threats in broad themes and presents a laundry list of unprioritized priority actions for departments and agencies to prioritize in their own planning. Notwithstanding the NSS process, the State Department approaches policy planning less methodically than does the Defense Department or other agencies in the national security enterprise.[9]

Indeed, in recent decades the State Department has instituted a number of different policy-planning processes intended to guide the implementation of diplomacy and foreign assistance.[10]

The Government Performance and Results Act (GPRA) of 1993 and the GPRA Modernization Act (GPRMA) of 2010 require executive branch agencies to develop four-year strategic plans no later than February of the year immediately following a presidential inauguration.[11] Consistent with these acts, several recent Secretaries of State have put in place their own signature reform initiatives to prod the State Department to adopt a more coordinated and streamlined strategy-making process: Secretary of State Colin Powell's 2003 Diplomatic Readiness Initiative (DRI) initiated a joint strategic planning process with USAID. Secretary of State Condoleeza Rice's 2006 Transformational Diplomacy initiative created a "Director of Foreign Assistance" position to oversee US foreign aid and de-fragment US foreign assistance provision. Secretary of State Hillary Clinton's 2009 Quadrennial Diplomacy and Development Review (QDDR) process was modeled after the Pentagon's Quadrennial Defense Review (QDR) to better align the strategic planning and resourcing processes of the State Department, USAID, and DoD.[12]

In 2016, Secretary of State Rex Tillerson suspended the QDDR process and instead directed the Department of State and USAID to develop a State-USAID Joint Strategic Plan (JSP). The FY2018-2022 JSP outlined four strategic goals (protect American security, renew America's competitive advantage, promote American leadership, and ensure accountability to the taxpayer), with three to five subsidiary objectives for each goal and two to three performance goals for each objective—fifty performance goals in total, largely qualitative rather than quantitative.[13]

USAID is "an independent establishment in the executive branch" by statute, but the USAID administrator works "under the direct authority and foreign policy guidance of the Secretary of State" according to 1998 law, and USAID's budget is overseen by the State Department to ensure coordination with broader US foreign policy objectives.[14] USAID and the State Department have complementary missions with different goals that advance America's interests around the world. USAID is focused on "ending extreme poverty and promoting resilient, democratic societies," while the State Department's mission is to "shape and sustain a peaceful, prosperous, just and democratic world."[15]

The different missions and cultures of the State Department and USAID can sometimes lead to bureaucratic tension in coordinating policy and program priorities. The State Department is a highly centralized, hierarchical organization that prioritizes caution over efficiency or speed of impact. USAID, in contrast, operates more entrepreneurially, focusing on how to maximize the impact of development programs on the ground and accepting the risks of failure as it iterates and adapts its programs quickly to changing conditions.[16]

The organizational structure of the State Department reflects the duality of US foreign policy goals and priorities. Many objectives are necessarily country- and region-specific, which requires regional expertise as represented by State Department's seven regional bureaus, which report to the Under Secretary for Political Affairs. Foreign policy priorities that are more global or transnational—such as economic and business affairs, counterterrorism, nuclear nonproliferation, human rights, refugees and migration, environmental issues, etc.—are managed by functional bureaus specializing in those issues, which report to Under Secretaries for Economic Policy, Arms Control and International Security, or Civilian Security and Human Rights. However, because most diplomacy is conducted at the bilateral level by US missions and embassies or multilaterally at the United Nations (UN) and regionally based organizations, and because regional bureaus control the budgets and staffing assignments to US embassies, regional bureaus tend to exert more influence over both policy and budgetary prioritization than functional bureaus do.[17]

Informed by the Joint Strategic Plan, the State Department's regional bureaus develop Joint Regional Strategies (JRS) while the State Department's functional bureaus develop Functional Bureau Strategies (FBS). These bureau strategies present the goals and objectives that US diplomacy is expected to achieve globally and US missions in each region. However, like the higher-level strategies, these bureau-level strategies do not address the resource requirements necessary to achieve the stated goals. The primary purpose of these strategies is to guide US missions and embassies as they develop their own four-year Integrated Country Strategies (ICS), the final and most operational stage of the foreign policy–planning process.

Integrated Country Strategies (ICS) are intended to articulate the US policy priorities in every country. The process of drafting the ICS is led by the Chief of Mission or the Deputy Chief of Mission. In theory the global and regional goals, objectives, and priorities described in all higher-level strategies are reflected in the ICS, where they are tailored to express the best ways to achieve those goals and objectives based on conditions in each country.

However, like a game of long-distance "telephone" played out over an 18-month cycle, often the objectives described in an ICS do not fully match the overarching goals described in the higher-level strategies. Unlike the higher-level strategies, the ICS is the only document in the policy-planning process that begins to address resource needs and budget requirements, though it does so in narrative form, qualitatively rather than quantitively. This is the only point in the policy-planning process where a direct linkage to resource planning is established, with the expectation that an approved ICS will help frame and inform every US mission and embassy's annual mission resource request, described next.

BUDGET AND APPROPRIATIONS PROCESS

The current process for determining and enacting the foreign affairs budget is dominated by the Department of State, Foreign Operations, and Related Programs (SFOPS) appropriations legislation, which includes State Department diplomatic and administrative operations (Title I of SFOPS appropriations) as well as foreign assistance (Titles II–VI of SFOPS appropriations); SFOPS captures 97 percent of discretionary funding in the international affairs budget.[18]

Foreign assistance is the largest component of the international affairs budget: in FY2020, US foreign assistance totaled $40.7 billion, or slightly under 1% of total US federal budget authority, disbursed to 180 countries, with Afghanistan, Israel, Jordan, Egypt, and Iraq receiving the largest amounts of US assistance.[19] When used effectively, foreign assistance can serve as a "force multiplier" to diplomacy, by strengthening friendly governance systems, building economic and commercial capacity to grow markets and jobs for US companies, bolstering the military readiness of US allies, and providing basic humanitarian assistance to vulnerable populations. By contrast, the State Department's enacted FY2020 diplomatic, consular, and administrative operations (D&CP) budget within SFOPS, which funds all diplomatic programs and salaries, embassy security, public diplomacy, educational and cultural exchange programs, State Department IT and human resources, and contributions to international organizations, totaled $17.3 billion.

The Foreign Assistance Act (FAA) of 1961 serves as the cornerstone for foreign affairs budget and assistance programs. The FAA merged the State Department's operating budget with foreign assistance funding, reformed the structure of foreign assistance programs, and established the process by which Congress authorizes and appropriates foreign affairs funding. The FAA required Congress to authorize broad funding targets for foreign assistance programs and then, in a separate subsequent legislative step, to appropriate specific funds to the State Department and other agencies.

In 1974, Congress created the international affairs budget function (Function 150) to combine, for accounting purposes, all foreign assistance programs implemented by any executive branch agencies to simplify Congress's oversight role, but at the same time Congress continued to legislate each agency's annual budget submission separately. Through 1985, Congress regularly enacted new authorization legislation or amended the FAA to update authorization time frames and to incorporate new programs and authorities.

After 1986, however, Congress turned more frequently to enacting freestanding authorities, rather than amending the FAA, and waived the requirement to authorize funds before making them available via appropriations legislation. The last time Congress passed a State Department authorization bill was FY2003.[20] Since then, Congress's annual SFOPS appropriations bill has been the primary legislative vehicle for funding the foreign affairs budget.[21]

In the absence of regular enactment of foreign aid authorization bills, the annual SFOPS appropriation bill has assumed greater significance for Congress in shaping US foreign policy resourcing. Not only do SFOPS appropriations bills set spending levels every year for nearly every foreign assistance account as well as the State Department's operations accounts but also establish new policy initiatives that would otherwise be debated and enacted as part of authorizing legislation.[22]

Developing, negotiating, approving, and enacting the annual SFOPS bill is an unwieldy process. The SFOPS appropriations process begins eighteen months ahead of a fiscal year's start on October 1 of a given year, with the State Department's Under Secretary for Political Affairs and Under Secretary for Management jointly issuing a mission/budget Resource request, instructing every State Department domestic bureau and US mission abroad—referred to as "operating units," or OUs—to develop an explanation of its strategic priorities for the coming fiscal year and the financial resources needed to achieve them.

On receipt of those instructions, over 200 OUs (domestic bureaus and overseas embassies and missions) develop unit-level budget proposals based on each unit's assessment of goals, objectives, threats, and resources, generally basing their policy priorities and resource requests on the prior year's priorities and budget as a baseline and marrying those quantitative figures with the qualitative narratives taken from the mission's or embassy's Integrated Country Strategy (ICS). The head of every OU—an ambassador or Deputy Chief of Mission at a US Embassy, or an Assistant Secretary (A/S) or Deputy Assistant Secretary (DAS) within a bureau

—are responsible for ensuring that the operating unit's priorities and resource requests are as consistent as possible with the National Security Strategy and the State-USAID Joint Strategic Plan.

Operating unit-level budget proposals are then reviewed by each bureaus' policy and budget offices: the Office of Foreign Assistance Resources, which oversees foreign assistance programming; the Budget and Planning Bureau, which administers funds for diplomatic activities like representation, reporting, and negotiations; and finally, the White House's Office of Management and Budget (OMB).

The OMB reviews each operating unit's request for alignment with the president's priorities and anticipated funding levels and will often send the proposals back to operating units for revision, a negotiating process known as "passback." Once that stage is complete, typically in February some eight months before the start of the fiscal year being budgeted, the OMB submits the president's budget to Congress at the same time that the State Department submits its agency-wide budget proposal, referred to the Congressional Budget Justification (CBJ). Several weeks after submitting the CBJ, the State Department will typically submit annexes providing detailed country-level account breakdowns.[23]

With the State Department's CBJ and country annexes in hand, Congress holds budget hearings with senior State Department officials called to explain and defend the budget request, and then formulates annual SFOPS appropriations. Under normal circumstances, Congress approves the SFOPS appropriations bill before the start of the fiscal year. Congress often uses the SFOPS legislation to depart from what the State Department requested in the CBJ and direct the State Department to expend specific funds on certain programs or countries based on congressional preferences.[24] These are described in allocation tables appended to the SFOPS bill.

This triggers a requirement by the State Department to submit to Congress a follow-up report—the 635(a) report—demonstrating that it is allotting funds consistent with Congress's specific funding directions and allocation tables. If the State Department's operating units wish to transfer or reprogram funds from one country or program to another, the State Department must notify Congress to request approval. After all of these steps are complete, operating units may finally obligate and spend the funds that have been allotted to them, some eighteen months after the original funding request was prepared.

THE CHALLENGES TO INTEGRATING ENDS, WAYS, AND MEANS ON FOREIGN AFFAIRS

Any critique of the State Department's approach to planning and resourcing US foreign policy must begin by recognizing the challenges imposed by the bifurcated processes used to conduct planning and resourcing. "It is extraordinarily rare for a coherent strategy to be produced by a bureaucracy of any sort, rarer still for it to emerge from the contest between the executive and legislative branches of government."[25]

Both processes are hamstrung by fragmentation and complexity that neither the top-down, NSC-led national security planning process, nor congressional oversight of the foreign affairs appropriations process, nor periodic attempts by newly arrived Secretaries of State to impose their own approaches to policy planning have been able to resolve. With twenty-five separate agencies in the mix, there are insufficient mechanisms and safeguards in place in Congress, at the NSC, and within the State Department to ensure ends, ways, and means are integrated and rationalized across the interagency and within agencies. We can categorize these systemic obstacles as falling within five spheres: congressional, interagency, organizational (within the State Department), cultural (within the State Department), and programmatic (within the State Department).

Some observers argue that Congress's abandonment of the foreign affairs authorization process in favor of reliance on the appropriations process to approve foreign affairs funds and direct how they are spent at the country level has allowed short-sighted politics and parochial horse-trading to shape the SFOPS budget. Any hope of applying a more coherent and strategic logic to resourcing American statecraft must start with Congress and should include measures to strengthen the foreign policy committees and restore the practice of annual foreign policy authorization bills.

If Congress were to recommit to the intentions of the 1961 FAA by debating and adopting comprehensive foreign affairs reauthorization legislation, inviting the heads of all executive branch departments and agencies in the national security enterprise to testify together, this could facilitate a more strategic and integrated review of US foreign policy priorities, connect the State Department's foreign policy priorities more directly to other national security priorities, and engage the American public's attention more effectively than the current, fragmented appropriations process allows for.

A related obstacle is that foreign affairs resourcing is a subordinate priority in the eyes of Congress and a national security enterprise tilted heavily toward the military instrument of power. As numerous former senior defense officials have cautioned, US foreign policy is dominated by the military, both in terms of the Pentagon's budget (the FY2020 defense budget was $718 million, about thirteen times larger than the entire SFOPS enacted budget of $54.8 billion and forty-two times larger than the $17.3 billion appropriated to State Department operations and diplomatic engagements), and in terms of political prioritization from Congress. [26] In his article for *National Interest*, Charles Stevenson noted:

> Defense wins out because of public support for the US military, for the jobs it creates across the country, for the threats that seem most significant. The US military ranks at the top of US governmental institutions in terms of prestige and public approval, where it has been for many years. Spending for the State Department and other civilian foreign policy authorities is so unpopular in Congress that

it hasn't passed a foreign aid authorization bill since 1986 or a State Department authorization bill since 2003.[27]

Moreover, when Congress enacts budget deals, it often combines international affairs with domestic program budgets while defense spending is put in a category of its own, as was the case in both the Budget Control Act of 2011 and the Budget Enforcement Act of 1990. To rectify this, Congress could consider dissolving the separate categories of defense and foreign affairs spending and merge the military and nonmilitary authorization processes into one comprehensive national security budget-authorization process.

A national security budget-authorization process could better balance tradeoffs exclusively within national security priorities rather than subject the foreign affairs budget to congressional horse-trading with domestic programs. It would also put lawmakers who understand foreign affairs policies and programs at the same committee table with colleagues who focus on the military and would bring generals and ambassadors to the same table at authorization hearings to defend their budget requests in more coordinated fashion.

The executive branch must also approach national security strategy-making and resourcing more holistically. With the State Department's lead role on diplomacy and development, the benefits of including diplomatic resourcing in a national security budget is self-evident. Bringing all national security agencies to the table under White House and NSC coordination would promote more effective balancing of competing priorities, more efficient allocation of resources, and better integration of all tools of statecraft. In contrast to conventional interagency and inter-budget committee squabbles, such an effort would be more likely to produce a genuinely strategic and widely supported national security strategy.[28]

Within the State Department, a range of impediments to more effective synthesis of foreign policy ends and means exist at the organizational, cultural, and programmatic levels. First, the State Department's policy-planning and resourcing processes run largely independently of each other, managed by different actors, under different authorities, on different timelines, with no formal linkages between the two. The linkages that do exist depend on the individuals responsible for drafting annual mission and bureau resource requests understanding the importance of tying resource requests to (often-outdated) strategy documents as best as they can.

If diplomatic goals have evolved since those strategy documents were issued, which is not unusual in regions facing conflict or instability, it falls on each operating unit to understand what the State Department's latest policy goals are, to request corresponding resources, and to hope that by the time the eighteen-month resource process is complete, their requests will still align with current diplomatic goals. Meanwhile, every resource request, upon reaching the Under Secretary, becomes subject to intra-bureau haggling and senior-level reprioritization, and then once submitted to OMB becomes vulnerable to shifting guidance from the White House and the OMB to ensure alignment with the president's latest priorities.

To fix this disconnect, the State Department should reorient its foreign policy–planning process to occur annually, concurrently with the annual bureau and mission resource request process. A unified combination of policy- and resource-planning process could marry the bottom-up process of requesting resources country-by-country level with the top-down process of setting global and regional policy priorities. The crucial task of balancing tradeoffs and integrating ends and means should happen at the Under Secretary level, by a unit of properly trained strategists working jointly for the Under Secretary for Policy and the Under Secretary for Management. This integration unit should be granted the sole function of shepherding this unified strategy-making process on a continuous basis, constantly evaluating changing needs on the ground, weighed against quantitative measures of effectiveness of specific programs and engagements, with the authority to adjust both ends and means without waiting for the next annual planning process.

Implementing this recommendation would most likely require a complementary effort to rationalize the current patchwork of authorities which govern the State Department's oversight of foreign policy and foreign assistance planning and implementation. These authorities are described in the Foreign Affairs Manual (FAM), which is intended to serve as the comprehensive, authoritative source for the State Department's organization structures, policies, and procedures.[29]

In brief, the legal authorities and administrative directives which govern the formulation and conduct of foreign policy are derived from the US Constitution as well as numerous statutes and executive orders, including the State Department Basic Authorities Act of 1980, 70 Stat. 890, as amended, the Foreign Service Act of 1980, Public Law 96-465, as amended, and the Omnibus Diplomatic Security and Antiterrorism Act of 1986 (Public Law 99-399).[30]

Authority for implementing foreign policy on a regional or bilateral basis is delegated through the Secretary of State to the Under Secretary for Political Affairs and to the six geographic bureaus.[31] In addition, US ambassadors accredited to a specific country receive a letter of instruction from the president which provides them full authority to oversee both policy implementation and resource and management responsibilities for the US diplomatic presence in that country.[32]

A separate delegation of authority from the Secretary of State assigns authority for the formulation and implementation of cross-cutting "functional" policies (e.g., human rights, arms control, refugee issues, counterterrorism; international narcotics, and law enforcement) to the appropriate Under Secretary among the four functional bureau "families."[33]

Finally, another delegation of authority from the Secretary of State assigns authority for overseeing all resource, budget, and management functions and processes to the Under Secretary for Management and to the Bureau of Budget and Planning to "direct the planning, development, and conduct of the Department's integrated planning, performance, and budget processes."[34] Any reform of these policy- and resource-planning processes should begin by simplifying these dispersed and often-competing delegations of authority, with the aim of establishing a specific authority to integrate the policy-planning and resourcing processes at the Under-Secretary level.

Even such a unified planning process, however, would still face hurdles within the State Department's dominant bureaucratic culture of the Foreign Service, made up of some 8,000 commissioned Foreign Service Officers (FSO) whose primary mission is to serve as American diplomats overseas. [35] An FSO is expected serve the bulk of a career abroad in at least two regions and to develop political, economic, and linguistic expertise in those regions. Most FSOs receive extensive training in foreign languages, regional studies, and diplomatic tradecraft, but very few have prior background in or the opportunity to develop expertise in national security strategy-making, foreign assistance program planning, budgeting, implementation, or monitoring and evaluation. Given the importance of regional expertise to career advancement, it is not considered professionally beneficial for most FSOs to serve in offices that specialize in resource and budget issues. Moreover, unlike the military, the Foreign Service does not emphasize strategic resourcing as a core element of foreign policy making, and does not encourage long-term training in leadership, management, or strategy-development beyond a one-week mandatory leadership training class after an officer is promoted —on average every five to seven years.

On-the-job experience is the primary teaching vehicle that prepares officers for advancement to the Senior Foreign Service (SFS), which can leave officers unprepared for the challenges of applying strategic logic or tradecraft when called upon to prepare either policy-planning strategies (the ICS) or resource planning documents (mission and bureau resource requests).[36] To rectify this gap, the State Department should provide mandatory training specifically relating to national security planning and resourcing for all FSOs at the junior, middle, and senior levels. This training could be designed uniquely to FSO needs or could be integrated with similar courses taught in the Capstone and Pinnacle programs at the National Defense University (NDU) and DoD's service-based war colleges.

Instituting this would require changes to the State Department's personnel system to ensure that long-term training unrelated to an onward assignment is considered career-enhancing and does not disadvantage an FSO in the eyes of annual promotion panels. To ensure agency-wide buy-in, FSOs should also be required to serve at least one assignment in a position focused on foreign policy planning and resourcing as a prerequisite for promotion into the Senior Foreign Service.

A related obstacle to the Foreign Service's cultural challenge is the lack of a standardized methodology to measure performance impacts and effects of diplomatic engagements and foreign assistance programs. In most cases, it is difficult to measure the success of a diplomatic engagement or foreign aid program. Current evaluation practices focus on whether an event or program took place within budget, not what the lasting outcomes and impacts were. Moreover, bureaus are not required to measure policy or program performance against their Joint Regional Strategy (JRS) or Integrated Country Strategies (ICS) goals and objectives.

The lack of formal monitoring and evaluation metrics to measure progress, especially towards goals that require long-term attention such as economic reform or democracy promotion, also diminish an embassy's ability to make program or resource adjustments as conditions evolve.[37] The absence of effective outcome monitoring prevents the resource request process of missions and bureaus from generating valuable insights about which engagements and programs are worth amplifying and which should be discontinued—critical information for an effective foreign policy–planning process.

At the time of this writing, the State Department has begun a National Interest Global Presence project to develop new quantitative metrics to analyze funding for diplomatic engagements and compare them with US national interests and policy priorities. The State Department is implementing this project through its new Center for Analytics tasked with implementing agency-wide data management and cutting-edge analytical tools.[38] This effort holds promise.

CONCLUSION

After twenty-five years of American primacy following the fall of the Soviet Union, the emergence of renewed great power competition in a messy multipolar world is forcing the US government to comprehensively reassess threats and opportunities to national interests. In such a dynamic and dangerous setting, the design and conduct of effective diplomacy, foreign affairs, and foreign assistance take on ever greater urgency.

As currently constructed, the processes for formulating and resourcing US foreign policy are sufficient to muddle through with diplomatic tools and programs that remain planned and funded well enough to react to global developments in ways that protect America's vital national interests. However, since policy failure is most often the result of poor planning or poorly managed implementation or both, muddling through comes at a cost.

The US government is at risk of missing an urgent opportunity to reform the architecture of foreign policy planning and resourcing in potentially transformative ways, from Congress to the NSC to the State Department's organization and culture. Given the speed and complexity of political, economic, social, and especially technological change in the world today, the risk of a new and even more dangerous global security paradigm emerging in the coming decade that could pose not just serious but existential threats to core US interests cannot be ruled out. The time to optimize the formulation and resourcing of American diplomacy and foreign assistance is now.

NOTES

1. Senate Armed Services Committee, "Defense Authorization for Central Command and Special Operations Command," CSPAN, March 5, 2013, https://www.c-span.org/video/?c4658822/user-clip-mattis-ammunition.
2. Deibel, *Foreign Affairs Strategy*, 209.
3. National Constitution Center, "Happy Birthday to the Department of State," Constitution Daily (National Constitution Center, July 27, 2020), https://constitutioncenter.org/blog/happy-237th-birthday-to-the-department-of-state.
4. Office of the Historian, Foreign Service Institute, U.S. Department of State, "Department of State History," https://history.state.gov/departmenthistory.
5. U.S. Department of State, "Duties of the Secretary of State," https://www.state.gov/duties-of-the-secretary-of-state/.
6. Adams and Williams, *Buying National Security*, 11.
7. Whitaker et al., *The National Security Policy Process*, 4.
8. The White House, "Memorandum on Renewing the National Security Council System," February 4, 2021, https://www.whitehouse.gov/briefing-room/statements-releases/2021/02/04/memorandum-renewing-the-national-security-council-system/.
9. Mazaar et al., *The U.S. Department of Defense's Planning Process*, 7.
10. Pramanik, "Reforming Diplomacy."
11. U.S. Department of State, "Foreign Assistance Resource Library," https://www.state.gov/foreign-assistance-resource-library/#planning.
12. Pramanik, "Reforming Diplomacy."
13. U.S. Department of State and U.S. Agency for International Development, "Joint Strategic Plan FY2018-2020," https://www.state.gov/wp-content/uploads/2018/12/Joint-Strategic-Plan-FY-2018-2022.pdf
14. Pramanik, "The Case for Keeping USAID and the State Department Separate."
15. US Department of State, "About the State Department," https://www.state.gov/about/about-the-u-s-department-of-state/.
16. Andrew Natsios, "Tillerson wants to merge the State Dept. and USAID. That's a bad idea," editorial, *Washington Post*, June 28, 2017.
17. Beecroft and Naland, *Strengthening the Department of State*.

18. U.S. Library of Congress, Congressional Research Service, *The Foreign Assistance Act of 1961* by Rennack and Chesser.

19. U.S. Agency for International Development, "Foreign Assistance Tracker," https://www.foreignassistance.gov/.

20. U.S. Government Printing Office; *Foreign Affairs Authorization Act, Fiscal Year 2003*, https://www.govinfo.gov/content/pkg/PLAW-107publ2 28/pdf/PLAW-107publ228.pdf.

21. U.S. Library of Congress, Congressional Research Service, *The Foreign Assistance Act of 1961* by Rennack and Chesser.

22. U.S. Library of Congress, Congressional Research Service, *Foreign Assistance* by Lawson and Morgenstern.

23. U.S. Library of Congress, Congressional Research Service, *U.S. Foreign Assistance* by Nick Brown.

24. Indeed, the Trump Administration has proposed deep cuts in the international affairs budget every year over the past three years, but Congress has not enacted those proposed cuts. Instead, Congress has maintained level increased funding for the foreign affairs budget.

25. Deibel, *Foreign Affairs Strategy*, 10.

26. U.S. Library of Congress, Congressional Research Service, *Department of State, Foreign Operations, and Related Programs* by Gill, Lawson, and Morgenstern.

27. Charles Stevenson, "How the NDAA Helps Militarize American Foreign Policy," *The National Interest*, July 27, 2020, https://nationalinterest.org/blog/buzz/how-ndaa-helps-militarize-american-foreign-policy-165600.

28. Bedford, Beers, et al., *State Department Reform Report*.

29. U.S. Department of State Foreign Affairs Manual and Handbook, https://fam.state.gov/Fam/FAM.aspx?ID=01FAM.

30. U.S. Department of State Foreign Affairs Manual and Handbook, "1 FAM 012: The Secretary of State's Authority," https://fam.state.gov/FAM/0 1FAM/01FAM0010.html.

31. U.S. Department of State Foreign Affairs Manual and Handbook, "1 FAM 100: Geographic Bureaus," https://fam.state.gov/FAM/01FAM/01 FAM0110.html.

32. U.S. Department of State Foreign Affairs Manual and Handbook, "2 FAM 113, Mission Functional Responsibilities," https://fam.state.gov/FAM/0 2FAM/02FAM0110.html.

33. U.S. Department of State Foreign Affairs Manual and Handbook, "1 FAM 200, Internal Functional Bureaus," https://fam.state.gov/Fam/FAM.aspx?ID=01FAM.

34. U.S. Department of State Foreign Affairs Manual and Handbook, "1 FAM 620 Bureau of Budgeting and Planning (BP)," https://fam.state.gov/FAM/ 01FAM/01FAM0620.html.

35. "Global Talent Management Fact Sheet," U.S. Department of State, June 30, 2020, http://www.afsa.org/sites/default/files/0620_state_dept_hr_ factsheet.pdf.

36. Bedford, Beers, et al., *State Department Reform Report*.

37. "Inspection of the Bureau of Near Eastern Affairs," Office of the Inspector general (OIG), U.S. Department of State, May 2017, https://www.stateoig. gov/system/files/isp-i-17-22.pdf.

38. U.S. Library of Congress, Congressional Research Service, *U.S. Overseas Diplomatic Presence* by Gill and Collins-Chase.

CHAPTER 6

RESOURCING
PARTNERS AND ALLIES

THE UNITED NATIONS

Rebecca Patterson

Not only is UN peacekeeping a cost-effective alternative to putting
our own soldiers in harm's way, it works!
—Lieutenant General John Castellaw (retired),
United States Marine Corps[1]

INTRODUCTION

As of February 2021, the United Nations (UN) operates twelve peacekeeping missions worldwide, with more than 88,000 military, police,
and civilian personnel. The United States is the single largest financial
contributor to United Nations peacekeeping activities—spending between
$1–2 billion per year and playing a key role in establishing, renewing, and
funding them. Both Congress and the executive branch shape US policy
and strategy towards the UN. Congress authorizes, appropriates, and

oversees US funding to the UN, while the executive branch represents the US in UN bodies through the State Department and the US Mission to the UN.[2]

The formulation of policy on UN peacekeeping over the last seventy years has largely been evolutionary and ad hoc. Peacekeeping was not regarded as a primary function of the UN when the organization was established in 1945. Indeed, there is no reference to the term "peacekeeping" in the charter. Yet the deployment of uniformed and civilian personnel to observe and monitor peace and ceasefire agreements quickly became a valuable tool for protecting international peace.

Since 1948, the processes that have guided the authorization, operationalization, financing, and professionalization of peacekeeping operations have evolved both in terms of the formal processes and procedures as well as the norms that guide how states and their diplomats interact at the UN. Many of the mechanisms and practices that have developed have been prompted by a range of crises resulting from the deployment and management of UN peacekeeping operations.[3]

The UN Security Council (UNSC), a 15-member organization[4] in which the United States has arguably the largest voice, both as a product of its role in creating the UN as well as its role as the largest financial contributor, authorizes what peacekeeping missions should be undertaken, how big the missions will be in terms of uniformed force size, and what tasks the peacekeepers should undertake.

However, the UN General Assembly, an organization comprising 193 members, in which each state has an equal voice, approves the budget for each mission independently. Despite the structural challenges that are built into the system, somehow the UN peacekeeping enterprise still works relatively well in terms of overall outcomes for international peace and security. In examining its effectiveness over the last sixty years, peacekeepers demonstrably reduce the likelihood that war recurs,[5] reduce the likelihood an internal conflict spreads outside a nation's borders,[6] contribute to regional stability,[7] reduce the length of conflict,[8] reduce violence, and increase post-conflict democratization.[9]

The primary obstacle to matching the ends and means of strategy for UN peace operations is by design—structural in nature. To effectively advocate policy change, US personnel require a deep understanding of this structure, how the pieces fit and work together as well as when and where there is opportunity for change.

This chapter provides an overview of how the United Nations plans, budgets, and implements peace operations around the world with a focus on the role that the US government plays in the formulation of peace operations' mandates and the negotiation of their budgets. It highlights the unique challenges of working in a multilateral setting and concludes with recommendations for US national security professionals seeking to more effectively tune the UN's resources to its strategic priorities.

GENERATING PEACEKEEPING MANDATES

The UN Charter, a treaty ratified by the United States on August 8, 1945, gives the UNSC primary responsibility for the maintenance of international peace and security, which requires that the council be available for emergency action and response.[10] Peacekeeping has evolved into one of the main tools used by the United Nations to achieve this purpose.[11]

Peacekeepers are deployed on the basis of a mandate from the UNSC—usually as a resolution—and their tasks differ from situation to situation, depending on the nature of the conflict and the challenges it presents.[12] Although each operation is different, there is a considerable degree of consistency in the tasks assigned by the UNSC, such as deploying to prevent the outbreak of conflict or the spillover of conflict across borders, stabilizing conflict situations after a ceasefire, assist in implementing comprehensive peace agreements, and leading states through a transition to stable government based on democratic principles.

Peacekeepers are often mandated to play a central role in disarmament, demobilization and reintegration of ex-combatants, security sector reform, protection and promotion of human rights, support for restoration and extension of state authority, promotion of economic recovery and development, and electoral assistance. In addition to delineating and prioritizing peacekeeping tasks, mandates set forth the ceiling for uniformed personnel, establish the mission's duration, and enumerate reporting requirements to the UNSC.[13]

An essential precondition for the creation of a UN peacekeeping mission is the consent and cooperation of the parties in conflict, which grew out of the UN's first deployment to the newly formed state of Israel in 1948.[14] The deployed force is also intended to be removed from politics and to refrain from using force except in the case of self-defense.[15] These three elements—consent of the parties, impartiality, and non-use of force except in self-defense and defense of the mandate—are referred to as the UN's peacekeeping principles. These three foundational elements of any peacekeeping force stand in stark contrast to the situation in which the United States would deploy its military force. When the US military deploys, it tends to do so without consent of the related parties, in a very partial way and with the intent to use force to impose its will.

The UN does not have any formal authority to plan for potential peacekeeping missions without direction from the UNSC. Furthermore, the nature of the UN response is highly dependent on the politics of the UNSC, the consent of the state experiencing the conflict, and the political will of other members states to volunteer to participate in a possible mission. For these reasons, as a conflict develops, worsens, or approaches resolution, the UN is frequently involved in a number of formal and informal consultations to determine the best response by the international community.

These consultations involve a variety of actors depending on the situation, but often include parts of the UN, the potential host government and parties on the ground, key member states, including those states that might contribute troops and police to a peace operation and regional organizations. The UN Secretary General (SG) may request a strategic assessment to identify all options for UN engagement, which can also include engagement with stakeholders. Ideally, following such a strategic assessment, the SG will deploy a Technical Assessment Mission (TAM) to report on the general security, humanitarian, and political situation on the ground. The SG uses these assessments to make recommendations to the UNSC regarding the establishment of a peacekeeping mission.

If the UNSC determines that deploying a UN peace operation is the most appropriate step to take, it will formally authorize the mission by adopting a resolution, which sets out the mission's mandate and size and details the tasks it will be responsible for performing. The proposed budget and resources are then subject to General Assembly approval. Over time, mission mandates have grown significantly in scale and scope, such as the requirement to protect civilians, yet the resources committed to accomplish these tasks have not expanded at the same rate.

While the Permanent Five (P5) members generally set the parameters of peacekeeping missions, and the secretariat (the UN's administrative body headed by the Secretary General) may advise the council on the feasibility of deploying a mission in a certain security context, it is ultimately up to the entire UNSC whether to authorize the mission and its objectives.[16]

In April 2013, when the UN was considering how to respond to the crisis in Mali, the secretariat was pessimistic about deploying a mission, but in the end, the UNSC decided to deploy one anyway.[17] Eight years after the UN's deployment to Mali, the mission is considered to be the deadliest (240 fatalities[18]), with little progress made towards the tasks laid out by the UNSC.[19]

THE POWER OF THE PENHOLDER SYSTEM

Due to the ability of any P5 member (Russia, China, the United States, France, and the United Kingdom) to veto any resolution, the UNSC is unable to authorize peacekeeping missions when there is disunity among them. The United States is routinely frustrated by this process, most recently by P5 division over the Syria crisis and what to do about it.[20]

France, the UK, and the US (the P3) are usually the "penholders" for peacekeeping mandates, meaning that they guide the drafting process when missions are established and during their renewal.[21] The penholder system emerged in its current form in 2006, when the United States, the UK and France worked together on nonproliferation resolutions for North Korea and Iran.[22] The P3 have "penholder" status on the vast majority of issues that the UNSC considers, giving them the power to decide what, when, and how mandates are created and considered.

When "holding the pen," a permanent UNSC member usually decides what action the council should take and then drafts an outcome document (such as a resolution or presidential statement) that is negotiated with the other permanent members before sharing the text with elected members.[23] The penholder circulates a "zero draft," which forms the basis for the negotiations. For every peacekeeping mandate, but especially those that the US drafts, there is an entire interagency process underpinning the language as well as instructions for the US negotiators at the UN Mission in New York.

US INTERAGENCY PROCESS AND UN PEACEKEEPING MANDATES

Given that the typical timeframe for a mandate's expiration is one year, there is opportunity for a robust interagency process for assessing ongoing missions and strategizing how the UNSC might improve on its instructions to each mission. Each mandate's renewal date is embedded in it, providing transparency as to when and how the UNSC can make changes. Though the mandate will be formally addressed at particular intervals throughout the year based on the mandated reporting requirements as well as during the month of its expiration, if the situation on the ground changes dramatically for the worse (or better), the penholder could decide to draft a new mandate out of cycle.

Within the State Department's Bureau of International Organizations (IO) the Office of Peacekeeping is allotted specific funding from Congress to monitor and evaluate UN peacekeeping missions. Along with this funding comes the requirement to brief congressional staffers from relevant committees on a monthly basis about the progress or setbacks in every mission, as well as any monitoring and evaluation trips taken by IO personnel. Ideally such a trip would occur three to four months in advance of a mandate renewal. The findings of the trip would inform an interagency process of deliberation on what the US would like to see altered in the current mandate.

In cases where the US has the pen, alterations are more straightforward, though still subject to the nature of negotiations in the UNSC. When another P5 member holds the pen, both informal and formal consultations between the US and the penholder are instrumental in negotiating a favorable outcome. This process provides an opportunity to think strategically about each mission; however, the US interagency process combined with the complexities of multilateral negotiations complicate and dampen the possibility of significant change. The United States is excellent at developing coherent strategies for individual peacekeeping mission improvement. However, rarely does it examine the peacekeeping enterprise as a whole, looking into where there can or should be tradeoffs between missions.

The other opportunity for more than incremental change to a mandate is through the process of strategic reviews of the mission. The penholder, or any UNSC member, may seek to add language to a mandate that calls for a strategic review of the mission with a report by the Secretary General on its findings to be completed by a particular date. By mandating such a review, the UNSC can send a message that the mission or its mandate needs to be reexamined. These reviews often provide suggestions for the UNSC to consider that go beyond typical mandate renewal incrementalism.

Who Provides Peacekeepers?

The UN has no standing army, nor does it have its own police force. When the UNSC decides to initiate a new mission that requires force, the UN has to solicit member states for troop or police contributions. With the authorization to deploy force, the UNSC enumerates the ceiling on police and military personnel in the mission. Working with member states to identify and deploy personnel can take time—often more than six months to get boots and equipment on the ground.

Capability gaps of both personnel and equipment are almost constant features of UN peacekeeping operations. Such gaps can stem from the lack of a particular asset (shortfalls in aviation such as helicopters is common) but also the uneven performance of deployed units.[24] Though the UN has worked to formalize the force generation process, only time will tell if it is able to moderate the political nature of Troop Contributing Country (TCC) / Police Contributing Country (PCC) selection. [25] The US has worked to expand the supply of available peacekeepers for particular missions, urging states to pledge forces or unique capabilities to missions with shortfalls, and at times, even offering training or equipment to incentivize states to do so.

Who Pays for UN Peacekeeping Missions?

The UN Charter requires each member to contribute to the expenses of the organization, and those who fail to pay their assessed dues may have their voting rights revoked.[26] Peacekeeping expenses are a part of the UN's assessed contributions, which are required dues shared among UN member states. In the past decade, peacekeeping costs have escalated dramatically—in the 2000–2001 fiscal year, approved expenditures totaled $2.6 billion, but amounted to $6.5 billion for 2019–2020, reflecting the increasing size, complexity and logistical challenges of UN operations.[27] While peacekeeping costs dwarf the UN regular budget, $2.8 billion annually in 2019–2020, they remain very modest compared to global military expenditures ($1.9 trillion in 2019)[28] or other state expenditures; for example, the Beijing (2008) and the Sochi (2014) Olympics cost approximately $42.6 billion and $50 billion respectively.[29]

Since 1948, there have been battles over how missions should be funded. Differing views over the peacekeeping funding formula were addressed by applying an assessed scale, which was first applied in 1974.[30] Today, the UN apportions its peacekeeping expenses among member States according to a formulaic scale of assessment which broadly reflects states' gross national income, debt burden, and per capita income levels.[31] The peacekeeping scale modifies this slightly by providing discounts to developing states financed by premiums on the P5.[32]

In 2018, the United States' assessed share of UN peacekeeping costs was 28.4 percent, the P5 were jointly responsible for 54.0 percent, and developed states, which were not permanently on the UNSC, accounted for an additional 39.5 percent.[33] By contrast, the 156 UN members that identify as developing states were assessed for only 5.6 percent of peacekeeping expenses.[34] However, they furnished 90.1 percent of the UN's military peacekeepers as of June 2020 while the United States furnished 0.03 percent of this personnel and the P5 jointly accounted for 4.3 percent.[35]

The General Assembly adopts the peacekeeping scale of assessment every three years. For the United States to change its share of the overall peacekeeping budget, other countries have to pay more, making it one of the very few moments in diplomacy when there is truly a zero-sum game. The last time the United States overcame this hurdle was in the late 1990s, when Richard Holbrooke, then US ambassador to the UN, undertook a massive effort[36] that required dozens of diplomats working for more than a year, along with the support and commitment of both Congress[37] and the executive branch at the highest levels. Without that level of effort, substantial change in the US assessment rate is extremely unlikely.

Table 1. Top Financial vs. Top Troop Contributors to UN Peacekeeping Operations as of Feb 2021.

Top Financial and Troop/Police Contributors to UN peacekeeping Missions

Top Financial Contributors	Top Troop/Policy Contributors
1.United States	1. Ethiopia
2. China	2. Bangladesh
3. Japan	3. Rwanda
4. Germany	4. Nepal
5. United Kingdom	5. India
6. France	6. Pakistan
7. Italy	7. Egypt
8. Russian Federation	8. Indonesia
9. Canada	9. Ghana
10. Republic of Korea	10. China

Source: Adapted from "Troop and Police Contributors" and "How We are Funded" at www.un.peacekeeping.org/en

At the UN, this bifurcation of contributions creates conflicting interests and at times perverse incentives where peacekeeping financing and mission budgeting are concerned. Most peacekeeping expenditures arise from the costs associated with deploying uniformed personnel, largely reimbursing TCCs, or from operations requirements such as infrastructure and transportation for deployed personnel. Large financial contributors, in particular the United States, have an interest in limiting those expenditures. By contrast, the states furnishing most of the UN's uniformed peacekeepers stand to benefit from reimbursement rate increases and more expansively funded missions.[38]

THE UN'S PEACEKEEPING BUDGETARY PROCESS

Decisions about financing marginally influence the direction of UN peacekeeping. The UN General Assembly's Fifth Committee has responsibility for the UN's administrative and budgetary matters, including peacekeeping operations.[39] The Fifth Committee agrees on annual resolutions that are then adopted by the General Assembly to ensure the UN and its work continues to be funded. The committee meets during the session in May–June to consider the peacekeeping support account (which includes funding for positions at headquarters related to peacekeeping), the budget for shared service centers (such as logistics bases) and the budgets for most UN peacekeeping missions.[40] The budgets are usually agreed upon by the end of June to coincide with the July 1 start of the peacekeeping financial year.

The process of peacekeeping-operation budgeting begins a year before the start of the next fiscal year with the issuance of mission-budget instructions from the UN controller. For those missions already in progress, the Fifth Committee is only able to make marginal changes to the resourcing of personnel and equipment, making the initial allocation of resources all the more impactful. Interestingly, the budget process for missions is structurally divorced from the mandate-renewal process—which happens on a different timeline and is considered by the UNSC rather than the General Assembly.

The missions themselves are responsible for integrating what is happening on the ground with instructions from the UNSC, with the resources authorized by the General Assembly. Missions formulate their budget proposals and submit them to headquarters in December or January. The Advisory Committee on Administrative and Budgetary Questions (ACABQ), an expert committee of sixteen members (the United States typically has one position on the committee), holds hearings and issues recommendations to the Fifth Committee.[41]

The ACABQ wields significant weight in the negotiations, including the focus and direction of discussions within the Fifth Committee. Because it is meant to be a technical body, the ACABQ generally does not directly oppose the UNSC's approach to peacekeeping mandates but instead makes marginal changes to issues by modifying mission-provided recommendations. It can also elect to delve deeper into certain issues, pushing for more information or challenging the efficacy of requested personnel or equipment.

The Fifth Committee traditionally makes decisions by consensus. Its dynamics mirror those of the General Assembly with a divide between the countries in the Group of 77 (G77), a coalition of 134 developing countries and the major financial contributors to the UN. One of the key challenges is finding a compromise between the needs of missions in the field, member state interests in keeping particular positions or equipment, and the amount major financial contributors are willing to pay. Several major financial contributors, including the US, often predetermine what overall figure they are willing to accept for total requirements across all missions and related support budgets. This requires careful negotiation of a package deal that balances multiple issues and competing priorities.

The United States and Its Peacekeeping Contributions

Congress plays a key role in shaping US policy at the UN through funding and oversight. Each fiscal year, Congress authorizes and appropriates its contributions to peacekeeping. Despite the assessment rate levied by the United Nations, and its treaty obligations, peacekeeping contributions have been a political issue for the United States for decades. To that end, Congress capped US contributions at 25 percent,[42] regardless of the UN assessment rate (27.89 percent for 2020) and has enacted legislation linking funding to specific UN reforms benchmarks. At times, the US has effectively leveraged its financial position to push for peacekeeping reforms.

Regardless of the politics of the US contribution, the United States would have to pay significantly more to unilaterally field an intervention comparable to a UN mission. In 2007, the US Government Accountability Office (GAO) estimated that it would cost the United States eight times as much to conduct a multidimensional peacekeeping operation similar to the UN Stabilization Mission in Haiti (MINUSTAH).[43] In 2018, a similar inquiry by GAO found that it would have cost the US twice as much to conduct a multidimensional peacekeeping operation similar to the UN Multidimensional Integrated Stabilization Mission in the Central African Republic (MINUSCA).[44]

CONCLUSION

As US policy makers seek more effective and efficient peacekeeping missions, they should be aware of two key elements: *when* change is possible and *what* is the source of US leverage. The primary obstacle to a more rational alignment of ends and means for UN peacekeeping is the multilateral nature of the United Nations, which requires compromise between very divergent national interests and is a reflection of the politics of the international system itself. Nowhere is the division between states more apparent than between those that undertake the actual peacekeeping and those that pay for it. The US should carefully consider when UN peacekeeping is the appropriate policy tool and use its unique position on the UNSC to advocate strongly to constrain or encourage intervention depending on the circumstances.

In those instances where there is a clear logic for a UN peacekeeping intervention, the US has the opportunity to amplify its voice by taking the "pen" on the original mandate. The penholder has the greatest opportunity to impact mission size, its tasks (and therefore its budget) and reporting requirements—all factors essential to aligning means and ends. In instances where the US is not the penholder, there are still opportunities for changes in mission tasks, their prioritization, and the size of the uniformed force as the mandate renewal arises. The US has experienced mixed results with assessing and strategizing one mission at a time—it could be more impactful if the peacekeeping enterprise were considered holistically to better align the tradeoffs between missions when it comes to resourcing.

Given the US position as the largest financial contributor, it has an important voice in advocating for peacekeeping reform. At times, tying its funding to specific reform efforts has worked.[45] Regardless, the US has been most effective at pushing for strengthening peacekeeping when it actively engages in both the formal and informal processes of the UN by wielding its power and influence to generate positive change. [46] Paying its dues on time and in full adds to its legitimacy when it comes to such efforts. Rather than ignoring the US share of the peacekeeping bill, the best way to address assessment rate concerns is sustained engagement. As the window of opportunity for change only arises every three years, it is crucial to be prepared with a comprehensive negotiation strategy well in advance of the next cycle.

NOTES

1. John Castellaw, "UN Peacekeepers Help Keep Soldiers Safe," *US News and World Report*, February 23, 2018, https://www.usnews.com/opinion/articles/2018-02-23/un-peacekeepers-help-keep-us-soldiers-safe.
2. U.S. Library of Congress, Congressional Research Service, *U.S. Funding to the United Nations System*, by Blanchfield.
3. Sharland, "How Peacekeeping Policy Gets Made."
4. The UNSC is comprised of five permanent members (China, France, United Kingdom, Russian Federation, and the United States) and ten elected members which represent different regional groups: Three from Africa, three from Asia-Pacific, two from the Eastern European group, two from Latin American and Caribbean states, and five from Western Europe and others.
5. Fortna, *Does Peacekeeping Work*, 37.
6. Beardsley and Skrede Gleditsch, "Peacekeeping as Conflict Containment," 69.
7. Beardsley, "Peacekeeping and the Contagion of Armed Conflict," 1058.
8. Ruggeri, Dorussen, and Gizelis. "Winning the Peace Locally," 163.
9. Doyle and Sambanis, "International Peacebuilding," 782.
10. In the UN Charter, two articles establish this requirement: According to Article 24, "in order to ensure prompt and effective action," it is required by Article 28 to be "so organized as to be able to function continuously."
11. "Mandates and the Legal Basis for Peacekeeping," United Nations Peacekeeping, https://peacekeeping.un.org/en/mandates-and-legal-basis-peacekeeping.
12. For an example of a peacekeeping mandate resolution, see UNSC Resolution 2531 (2020), S/RES/2531(2020), https://undocs.org/en/S/RES/2531(2020).
13. When peacekeeping mandates are updated or reevaluated, especially if there is little change, a letter from President of the UNSC to the Secretary General, though not official, will be given the same weight and authority as a UNSC Resolution.
14. United Nations, "Fifty-Five Years of UNTSO," May 9, 2003, https://www.un.org/en/events/peacekeepersday/2003/docs/untso.htm.

15. See UN General Assembly, *Summary Study of the Experience Derived from the Establishment and Operation of the Force: Report of the Secretary General*. UN Doc. A/3943, October 9, 1958.

16. A resolution requires nine affirmative votes to be adopted.

17. See UN Security Council, *Report of the Secretary General on the Situation in Mali*, UN Doc, S/2013/189, March 26, 2013, https://undocs.org/S/2013/189.

18. The fatality data as of July 31, 2020. See https://peacekeeping.un.org/en/mission/minusma.

19. "Explanation of Vote on the Renewal of the Mandate for the UN Multidimensional Integrated Stabilization Mission in Mali (MINUSMA)," United States Mission to the United Nations, June 29, 2020, https://usun.usmission.gov/explanation-of-vote-on-the-renewal-of-the-mandate-for-the-un-multidimensional-integrated-stabilization-mission-in-mali-minusma/.

20. For just one example of many, see "Statement from Ambassador Craft on Russian and Chinese Vetoes of a UN Security Council Resolution to Extend Cross-Border Aid in Syria," by USUN Ambassador Kelly Craft, July 7, 2020, https://usun.usmission.gov/statement-by-ambassador-kelly-craft-on-the-russian-and-chinese-veto-of-the-syria-humanitarian-resolution/.

21. Loraine Sievers and Sam Daws, "Chapter 5: Conduct of Meetings and Participation" in "Section 6: Motions, proposals, and suggestions" in *The Procedure of the UNSC* (4th Edition) (Oxford: Oxford University Press, 2014), https://www.scprocedure.org/chapter-5-section-6c.

22. Loiselle, "The penholder system and the rule of law in the Security Council decision-making."

23. Ibid.

24. Smith and Boutellis, "Rethinking Force Generation."

25. UN DPO Guidelines, "Peacekeeping Capability Readiness System (PCRS)" January 1, 2019.

26. Article 19 of the UN Charter revokes the voting right of those member states in arrears.

27. UNSG, "Approved Budgetary Levels for Peacekeeping Operations for the Prior Period from 1 July 2000 to 30 June 2001," 10 August 2001, A/C.5/55/48; UNSG, "Approved Resources for Peacekeeping Operations for the Period from 1 July 2019 to 30 June 2020," 3 July 2019, A/C.5/73/21.

28. Nan Tian, Alexandra Kuimova, Diego Lopes Da Silva, Pieter D. Wezeman, and Simon T. Wezeman, "Trends in World Military Expenditure, 2019," Stockholm International Peace Research Institute (SIPRI) Fact Sheet, April

2020, https://www.sipri.org/publications/2020/sipri-fact-sheets/trends-world-military-expenditure-2019.

29. Paul Farhi, "Did the Winter Olympics in Sochi really cost $50 billion? A closer look at the numbers," *Washington Post*, February 10, 2014, https://www.washingtonpost.com/lifestyle/style/did-the-winter-olympics-in-sochi-really-cost-50-billion-a-closer-look-at-that-figure/2014/02/10/a29e37b4-9260-11e3-b46a-5a3d0d2130da_story.html.

30. Sandler, Todd. "International Peacekeeping Operations."*Journal of Conflict Resolution* 61, no. 9, (2017), 1878, https://doi.org/10.1177/0022002717708601.

31. Two peacekeeping missions deployed in the 1940s—the UN Truce Supervision Organization and UN Military Observer Group in India and Pakistan—are funded by the regular budget.

32. For a great explanation see Mark Leon Goldberg, "Three Things You Need to Know about the UN Scales of Assessment Negotiations Underway in New York," *UN Dispatch*, December 19, 2018. https://www.undispatch.com/three-things-you-need-to-know-about-the-un-scales-of-assessment-negotiations-underway-in-new-york/, and the UN General Assembly Resolution A/RES/73/372, January 3, 2019.

33. UN General Assembly, *Scale of the assessment for the apportionment of the expenses of United Nations peacekeeping operations*, A/73/350/Add.1, December 24, 2018, https://undocs.org/en/A/73/350/Add.1.

34. Ibid.

35. Data as of June 30, 2020; see https://peacekeeping.un.org/en/troop-and-police-contributors

36. Goldberg, "Three Things You Need to Know about the UN Scales of Assessment Negotiations Underway in New York."

37. The "Helms-Biden" accord, which constructed a package that called for partial payment of US arrears with subsequent payments predicated on lowering the US-assessed rate and hitting UN reform target, provided the framework for AMB Holbrooke to negotiate a reduction in US contributions to the regular budget. This had the knock-on effect of bringing down the peacekeeping share.

38. Coleman, "Extending UN Peacekeeping Financing Beyond UN Peacekeeping Operations?," 104.

39. UN Charter, Article 17, "The General Assembly shall consider and approve the budget of the Organization" and "The expenses of the Organization shall be borne by the Members as apportioned by the General Assembly." An advisory opinion by the International Court of Justice in

July 1962 found that peacekeeping was consistent with the aims of the organization and therefore fell under Article 17 as an assessed expense.

40. The two exceptions are the UN Truce Supervision Organization and the UN Military Observer Group in India and Pakistan, which are funded through the regular budget since they were both established before the creation of separate funding modalities for peacekeeping operations.

41. Members of the ACABQ are elected by the General Assembly for a period of three years on the basis of a broad geographical representation. Members serve in a personal capacity and not as representatives of member states. The committee holds three sessions a year with total meeting time between nine and ten months a year. See https://www.un.org/ga/acabq/about.

42. The 25% cap was enacted in the Foreign Relations Authorizations Act, Fiscal Years 1994 and 1995, April 30, 1994. For many years, Congress raised the cap in annual SFOPS acts. Since mid-2017, the Trump administration allowed for the applications of peacekeeping credits up to by not beyond 25%.

43. "Peacekeeping: Cost Comparison of Actual UN and Hypothetical Operations in Haiti," U.S. Government Accountability Office, February 2006, https://www.gao.gov/new.items/d06331.pdf

44. "UN Peacekeeping: Cost Estimate for Hypothetical U.S. Operation Exceeds Actual Costs for Comparable UN Operation," U.S. Government Accountability Office, February 6, 2018, https://www.gao.gov/products/GAO-18-243.

45. In particular, Congress has effectively pushed for UN reforms related to Sexual Exploitation and Abuse by UN peacekeepers by threatening to withhold funding.

46. Smith, "The United States of America," 45.

CHAPTER 7

THE DEFENSE BUDGET PROCESS

Tom McNaugher

The [executive branch agency] chiefs were constantly looking over
their shoulders...at the elements of the legislative establishment
relevant to their agencies—taking stock of moods and attitudes,
estimating reactions to contemplated decisions and actions, trying
to prevent misunderstandings and avoidable conflicts, and planning responses when storm warnings appeared on the horizon.

—Herbert Kaufman,
The Administrative Behavior of Federal Bureau Chiefs[1]

INTRODUCTION

The defining characteristic of the US budget process springs from the
defining characteristic of the US political system generally: separation
of powers. As the late Senator Daniel Patrick Moynihan put it some
years ago, "the United States is the only democratic government with a

legislative *branch*."[2] Applied to the budget process, the folks who provide the money—appropriations committees on both sides of Capitol Hill— lie separate from and independent of the folks who ask for it and spend it. It is a simple point with profound implications. Above all, there is no unity of command over the budget process, which makes it remarkably difficult to command things to happen.[3] Most of the time, policy makers move their initiatives forward by creating or exploiting a consensus across legislative and executive branches.

This consensus-building process is seen as taking place mainly at the highest levels, as say the president working with leaders in both houses of Congress to create each year's budget or some other major piece of legislation. But it occurs at all levels. Beneath the overall budget lie myriad cross-branch coalitions that sustain individual pieces of the budget. There is no requirement that these lower-level coalitions pull to the tune of the president's priorities or some overarching strategy document. And there are far too many of them to be managed individually. So, the overall budget is rarely, if ever, "rationalized" in the fullest sense of that term. Rather, each year's budget represents what American political scientists often call an "outcome"—the often-unintended result of tugging and hauling among bureaucrats and political players who have power and a stake in one or another outcome, and in the absence of an overarching authority that could seriously discipline the process.

If we look at the budget process specifically in the case of the Department of Defense, it has funded, over many decades, a remarkably effective military—many of the world's most sophisticated weapons and a highly trained professional force. It has done so, however, while also funding a great deal of duplication and inefficiency. For years, the United States has fielded two armies (the US Army and US Marine Corps) and three air forces (the US Air Force, Naval Air, USMC Air), or five if one includes US Army and US Marine rotary wing air forces. It fields only one capital ship fleet for the US Navy, although the Marine Corps' rather sizable fleet of amphibious ships might be considered a second, and the US Army fleet of harbor and logistics craft outnumbers the US Navy capital ships but serve a very different purpose.

At the programmatic level, the United States fields five fighter aircraft (F-15, F-16, F-22, F-18, and F-35), two separate fleets of helicopters (the US Army's Apache and Blackhawk), the Marine Corps' markedly different Sikorsky aircraft (plus a Super Cobra that dates from the Vietnam era), and the hybrid tiltrotor V-22. The US clings to new weapons long after rising costs have undermined their presumed cost-effectiveness. Despite several Base Realignment and Closure (BRAC) rounds, most analysts agree that the US still keeps more bases in operation than it needs. While democracies are by their nature inefficient—security dollars inevitably serve domestic political as well as security needs—this particular brand of democracy can be accused of going overboard!

Still, if Robert Gates is right when he opines on the nation's "perfect" record of predicting future crises—"we have never once gotten it right"—then there are serious questions about how "efficient" one wants the defense budget to be.[4] Surely it fails if it aligns closely with a published defense strategy like the National Security Strategy or its Pentagon derivative National Defense Strategy.[5] (Arguably published US strategy provides guidance to adversaries as well as the Department of Defense, suggesting what they should avoid doing.) The challenge would be less demanding if one force could handle all contingencies, as the services often claimed during the Cold War. But fourteen years of counterinsurgency in Iraq and Afghanistan have taught us that different operations require different forces and training. Efficiency under these conditions should be defined by hedging—focusing on top priorities, to be sure, but funding and fielding a diverse array of capabilities to better handle uncertainty and surprise. Diversity, rather than streamlining, should be the guiding motif for resourcing national security.

Arguably over many years the defense budget has given the nation more diversity than it needs. Could we reform the system in a way that pares down the diversity and redundancy to a more rational, analytically justifiable level? Constitutionally ordained separation of powers sharply limits attempts to discipline the interaction of the legislative and executive branches, and that in turn limits the effectiveness of reforms within each branch. End runs, or long-standing cross-branch coalitions, defeat the efforts of many reforms. As an example, in 1947 then US Army Chief of Staff, Dwight Eisenhower, suggested to members of Congress considering unification of the services in a "defense establishment" that it was time for the US Marine Corps to be folded into the US Army to create a unified US ground force. Today the Marine Corps is so solidly linked to key support on Capitol Hill that no serious defense official is willing to take on a major Marine Corps program, much less suggest eliminating the service itself. Although there have been plenty of reforms, the duplication and excess capacity mentioned earlier still exist.

This leaves the source of budget management and discipline to the political skills of managers within the security structure. This seems so obvious that it is a wonder how rarely it is observed in practice. Many still think the nation needs a successful businessman or woman to run the Defense Department, as if somehow running a private-sector business had much to do with overseeing a huge bureaucracy embedded in a complex political structure. Managing the defense-budget process is a highly political undertaking that requires a different set of skills—and even a skilled manager faces serious limitations.

THE FEDERAL BUDGET PROCESS IN OUTLINE

It is important to clarify at the outset that the budget process does not actually allocate most of the federal budget. Most of that budget goes to social welfare and support programs, notably social security, Medicare, and Medicaid, whose levels are mandated by law. These came to $2.7 trillion (61%) of the FY2019 federal budget, which totaled $4.4T. A separate and growing mandatory expense is the annual interest payment on the nation's deficit—$375B (just under 1%) of the FY2019 federal budget, and roughly half the size of the FY2019 defense budget. The budget process handles *discretionary* spending (about $1.3T in FY2019) which must be adjudicated annually within the political system. Security funding writ large is discretionary funding, and the defense budget alone consumes roughly half of the discretionary part of the budget (or about 15 percent of the federal budget). Given the relative dominance of the Defense Department among the nation's security organizations, this chapter will focus on it.

On paper, and in very simplified form, the budget process looks like two big circles that go out from and return to the president and his budget office, the Office of Management and Budget (OMB). Within the executive branch, the OMB provides executive branch agencies like the Defense Department with fiscal guidance—spending targets for the fiscal year in question. With a certain amount of help from the OMB, the agencies assemble their budgets within those targets, or targets renegotiated in the process, and return the finalized budgets to the OMB. The OMB puts the agency budgets (twelve separate budgets in all) together into an overall federal budget (the "PB" or "PresBud" for the President's Budget) and forwards it to Capitol Hill, normally on, but often only roughly close to, the required delivery date of the first Monday in February of the year preceding the fiscal year in question.

This launches the second circle, or rather two circles, since the president's budget is considered independently in the House and Senate before appropriators in each house meet in conference to finalize Appropriations Acts for executive branch agencies. It often looks like four circles since authorizing committees in both houses (for defense, the House and Senate Armed Services Committees) also consider the budget and produce National Defense Authorization Acts that look a lot like budgets, and these committees technically provide budget guidance to appropriators.[6] But no one can spend a penny on the basis of an authorization act; only an appropriations bill produces Budget Authority (BA), which is the authority to commit, or obligate, federal funds to executive branch undertakings. Both authorizations and appropriations acts go—

normally separately—to the White House for the president's signature, hopefully before the fiscal year begins on October 1. But Continuing Resolutions (CRs), which allow federal agencies to spend at the previous year's levels in the absence of a new appropriations bill, have been all too frequent over the past two decades.[7]

THE REALITIES: IN THE PENTAGON

The defense budget process as we know it today is relatively new in the context of the nation's history, the Department of Defense having come into existence in the years after World War II. For most of the nation's history it's two military services, the US Army and the US Navy, occupied separate cabinet departments (War and Navy), which planned and budgeted largely separately (despite the existence of an "Army-Navy Board") and worked independently with separate authorizing and appropriating committees and subcommittees on Capitol Hill. Even the White House was only marginally involved in these transactions; until Congress created the Bureau of the Budget (today's OMB) in 1921, executive branch cabinet departments submitted their budgets directly to the appropriate congressional organizations. With the experience of World War II suggesting that war had become a joint undertaking, the nation's politicians and senior executive branch officials debated unification of the services in the years immediately after the war and after much heated debate. created the defense establishment with a weak Secretary of Defense (initially just another cabinet secretary alongside the cabinet secretaries of War and the Navy) with the National Security Act of 1947.

Congress and the president set out almost immediately to strengthen the position of the Secretary of Defense, culminating in a major boost in the legal authority of the secretary and his staff, the Office of the Secretary of Defense (OSD) in the Defense Reorganization Act of 1958.[8] Robert McNamara fully leveraged these new legal powers to create the budget process we see in today's Department of Defense. McNamara implemented the Planning, Programming and Budgeting System (PPBS) in the first six months of the Kennedy administration as a means of linking strategy to budgets. Dismayed that the services had no systematic way of tracking the out-year implications of the current year's investments, McNamara created the Five-Year Defense Plan (FYDP) to project the budget year's numbers out four more years (the first year of the FYDP being the basis for his department's annual budget submission). Noting massive duplication of capabilities across the services, McNamara created nine Major Force Programs (MFPs), such as strategic forces, general purposes forces, and mobility forces, to group together for analysis similar capabilities from the different services. Finally, and most famously, McNamara deployed a team of mostly civilian systems analysts—the whiz kids, as they were called, often without affection—to perform rigorous analysis of trade-offs using quantitative data where possible.[9]

Neither McNamara nor his whiz kids are remembered with great fondness these days, and while their involvement in the nation's Vietnam fiasco has much to do with that, so did the seeming arrogance with which young whiz kids, often with no military experience, pushed aside the seasoned judgment of military professionals in favor of their own quantitative assessments. Yet in those first years McNamara succeeded impressively in shifting the focus of US military forces away from reliance on nuclear retaliation to a more robust conventional force that would offer the president options short of nuclear retaliation in a serious confrontation with the USSR. In part, this is because McNamara was implementing a new strategy, flexible response, to replace the Eisenhower administration's emphasis on massive retaliation—a strategy that seemed to a growing number of officials and observers to be dangerously outmoded in the face of Soviet nuclear and missile developments. The return to a more balanced conventional force posture, after years in which the US Air Force had pulled down nearly half the annual defense budget, certainly appealed to the US Army and the US Navy.

His success also reflected the sheer novelty of the new enterprise. In a sense, McNamara took the political system by surprise. Over his years in charge of the DoD, McNamara closed well over a hundred bases without much regard for the damage his actions inflicted on political districts.[10] He cancelled the B-70 bomber (ICBMs could invade Soviet airspace with much greater confidence) and the F-105 tactical aircraft, designed primarily to deliver nuclear weapons at the tactical as well as strategic levels.[11] Meanwhile, inside the Pentagon he took the Service Chiefs by surprise as well, playing them off against each other until the mid-1960s, when they realized that McNamara could be contained by the unanimity of the Joint Chiefs of Staff.

As controversial as McNamara became, both his management and strategic reforms have remained more or less the bedrock of the DoD's planning and procurement process to this day. While flexible response was created to handle problems in the Eisenhower administration's nuclear strategy, it gave the military services license to organize, train, and arm for conventional war. As the US commitment to Vietnam receded in the early 1970s, the services set about developing robust conventional forces, including weapons that in many cases are still around today, albeit in much modified and improved form. The US Army's M-1 tank, M-2 Bradley Fighting Vehicle, Apache attack helicopter, and Blackhawk utility helicopter, for example, all stem from requirements laid down in the early 1970s. The US Air Force's F-15E grew out of that service's early 1970s' F-15 program, while the air force's and navy's F-16 and F/A-18 fighter/attack aircraft were born in the Lightweight Fighter competition that same era. US military aircraft have moved beyond these technologies to 5th generation stealth with the B-2 bomber, the F-22 fighter, and the F-35 fighter/attack aircraft, but the B-2 entered development in the mid- to late-1970s.

Meanwhile, the PPBS remains the structure within which Pentagon agencies shape their budgets, although the "S" ("System") gave way to an "E" for "Execution" in 2003, in recognition of the fact that actual spending (i.e., budget execution), or failure to spend, did not always conform to the expectations of Pentagon officials or congressional appropriators. The services still produce FYDPs, estimating out-year costs (often optimistically) over five years, with the first year the basis for the upcoming budget submission. One can still find reference to McNamara's Major Force Programs, although even McNamara had problems making cross-service trades within these categories. Systems analysis also remains an analytical tool for examining major trade-offs, although other tools —notably "net assessment"—have come to occupy important positions in the Pentagon's analytical tool kit.

But while systems analysis remains important, its powerful role in the budget process has declined since the heady days of the McNamara era, when McNamara and his whiz kids virtually dominated the budget process, offering draft budgets to the services for comment.[12] This made the Office of Systems Analysis (OSA) controversial among senior uniformed officials as well as on Capitol Hill, especially among Republicans. Thus Melvin Laird, President Nixon's first Secretary of Defense (1969–1973) returned drafting power to the services:

> He revised the PPBS, including a return to the use of service budget ceilings and service programming of forces within these ceilings. The previously powerful systems analysis office could no longer initiate planning, only evaluate and review service proposals.[13]

Although different secretaries can use the services of DoD's systems analysts (since 2009 lodged in the Cost Assessment and Program Evaluation (CAPE) office) as much as they like, since Laird's time the services have generated their own budgets.[14] The OSD reviews these, and CAPE and the services home in on key issues requiring analysis during each budget cycle.

This brief summary cannot do justice to the incredibly labor-intensive and arcane bureaucratic processes that shape each year's defense budget. The process begins over a year before the budget in question is due on Capitol Hill early in February, almost two years before it becomes law—assuming the president signs the budget on time. Service budgets contain thousands of line items, and rarely does anything stay constant from year to year. Even putting aside congressional action or movement up or down of service top lines, most of the outyear numbers in each service's FYDP are optimistic, especially in the acquisition accounts. Under normal circumstances, fitting these programs into the next annual budget will require some adjustment, either stretching the program out or reducing funding elsewhere. Senior officials in the services or the OSD may have a priority they would like to see funded more generously going forward, and of course more funding to those line items will force compensating adjustments elsewhere in the budget. Needless to say, budget machinations get much easier if the DoD's top line rises, as it did, sharply, in President Ronald Reagan's first term. But this simply means that the defense budget is always pushing upward.

Despite all the sound and fury of the process, however, budget shares and programmatic content have been remarkably consistent over several decades. Taking out supplementals for war, for example, Tim Cooper and Russell Rumbaugh find that service budget shares have been remarkably consistent over four decades since 1973, even as "the US military went from Cold War to peace dividend to sustained irregular warfare during the war on terror."[15] Program Element (PE) funding may rise or fall from year to year, but programs largely remain funded. In his study of PPBS, retired US Army colonel Tom Davis noted that despite the enormous bureaucratic effort that goes into shaping each year's budget, "service programs as submitted to OSD following the programming phase of PPBS are little changed during the OSD review."[16]

Program stability reflects the consistency of service priorities over long periods, coupled with service control over initial budget drafts, courtesy of Secretary Laird. In a sense, power in the Pentagon flows down since civilians in OSD are reacting to complex budget drafts coming up from the services.[17] The consistency of budget shares and program content over so many years lends considerable credence to the long-heard adage that the first "P" ("Planning") in PPBS is silent. Strategy changes, but the forces do not change much. Secretary Robert Gates recognized this when he accused the services of spending too much time and money planning for some future wars while ignoring the two wars—in Iraq and Afghanistan—that their forces were fighting at the time.[18] In fact, during the decade after the 9/11 attack on the United States, the services used a substantial increase in defense spending to modernize their existing force structure.[19] Gates managed to force the Mine-Resistant Ambush Protected (MRAP) vehicles into the US Army and Marine Corps inventories only by using funds outside normal procurement channels, leaving service procurement accounts largely undisturbed.

ADD ON THE HILL

In theory, and to some extent in practice, the budget-producing process just described as well as the budget review and approval process on Capitol Hill are separate, sequential processes, neatly divided in time by the "drop" date in early February and in space by the Potomac River. After the budget drops, DoD officials, from the Secretary of Defense through the service secretaries and chiefs to individual program managers and key staff dutifully cross the river to defend the budget, or their portion of it, before the various committees and subcommittees that oversee defense. Each visit is normally preceded by staff exchanges that smooth the way for the hearing, avoiding surprises and identifying the key congressional concerns. It is an incredibly elaborate dance, mostly public, involving thousands of hours of testimony by hundreds of Pentagon officials to tens of committees and subcommittees. The process shines light into virtually every aspect of the Pentagon's budget.

No matter how carefully the budget and associated FYDP may have been assembled in the Pentagon, the budget will change on the Hill. DoD officials and members on both sides of the Hill will disagree about pay levels, recruiting issues, cost growth in weapons projects, readiness—

with the list going on and on. Congress will change, or occasionally even eliminate, line items in the submitted budget. This will of course force more changes to the budget and the FYDP, which mean more work for Pentagon staff.

Yet little of this will take Pentagon officials entirely by surprise. Few in the Pentagon are ignorant of what key members of congressional committees and subcommittees want to see in their area of concern about the defense budget. These officials may know what these committee members wish to see from last year's hearings, from public statements to the press, from personal visits to the Hill, or from perhaps thousands of brief staff exchanges over the year. The actual flow of a bill is not entirely predictable even to senior leaders on the Hill; it is how hundreds of members with differing opinions work out their differences that is difficult to predict. Ultimately, few in the Pentagon live in ignorance of what specific members think.

To the contrary, beneath the sequential and highly visible "formal" budget process lies a far less visible cross-branch dialogue involving officials and staff members from all levels, and that follows no particular schedule. Pentagon officials realize that members of Congress can make or break their budget, while also asking embarrassing questions in public hearings. This encourages them to reach out to those on the Hill who are important to their particular budget or program and to keep those individuals informed, invite their input, consult with them about changes, and carry on necessary exchanges. (Pentagon officials, even low-level staffers, can also reach out to the Hill with criticism of ongoing programs.) Even without visiting the Hill, Pentagon officials are likely to evaluate key defense policy decisions partly on the basis of the reaction those decisions will probably provoke on Capitol Hill. Good managers among the Pentagon officials—from the top on down to program managers—keep an eye on the Hill.

Often, of course, Pentagon officials do not have to reach out to members because members and their staff come to them with questions, complaints, and information. The members' interests may be grounded in pecuniary concerns—defense money flowing to their constituents, through DoD contracts for weapons and services, or, more importantly, the presence of military installations in districts. They may also be motivated by ideology or personal interest.[20] Member concerns rarely align with the timing of the budget process, nor does their understanding of the issues that concern them spring mainly from hearings. They or their staff visit installations and firms in their states and districts. They hear from firm lobbyists and post personnel. They may get calls from disgruntled service members or civilians unhappy with decisions affecting their concerns.

Such exchanges of information may be friendly, neutral, or acrimonious—but they tend to be very well-informed exchanges. Staffers on Capitol Hill often have followed key issues much longer than the DoD or service officials who come over to brief them have held their jobs.[21] Some staffers may have retired from Pentagon jobs and now follow their favorite issues from across the Potomac. And while members of Congress themselves may be stretched thinly across a wide range of issues they have to deal with each year, they or their staff are likely to be extremely well-informed about a post or installation in their state of district. They may visit the place on trips home and know some of the key people in charge. They may hear from post or installation personnel. They may have a better sense of what is going on at a particular post and how money appropriated to it is actually being spent—budget execution— than the budget personnel of the service in question.

This exchange of information may or may not be reflected in the budget that goes to the Hill usually in February. For one thing, the information coming over is often cacophonous, given how members usually have differing views on specific issues. Even if the message is fairly clear, Pentagon officials may decide to dispute the issue. Or they may game the issue, omitting an item that they know Congress will reinsert, hoping the Congress will also insert additional money (the Gold Watch syndrome[22]). Still, over the course of any fiscal year, many adjustments and even technical changes to developing weapons are put in place to head off confrontation and ease budget passage on the Hill.

Thus, a description of the DoD's PPBE system as it unfolds in the Pentagon—a description that leaves out Congress—fails to capture the realities and complexities of the process. Conversely, a perspective on Pentagon management that leaves the Hill out of the picture seriously underestimates the dimensions of the challenges DoD managers face. Dealing with members, visiting them to bolster support or head off trouble, and calculating the congressional reaction to Pentagon decisions goes on all the time.

And to what effect? Overall, congressional involvement in the budget process makes it that much more difficult to cancel anything, and this adds to the upward pressure on the budget. Bases, installations, and industrial contracts all bring money into states and districts that members of Congress would prefer to keep coming. To be sure, seismic global events can cause the budget to drop; for example, the end of the Cold War led to a decade's long effort to recoup the peace dividend.[23] Politicians are also sensitive to sunk costs, as are most members of the services. Acquisition projects that begin with optimistic cost estimates may thus enjoy continued support even as rising costs call into question the initial rationale for the system.[24] The long-term consistency of service procurement is due to Congress as well as service preferences.

But there are important caveats to this overall conclusion. Members of Congress can support innovation and innovators in the Department of Defense. Key members or committees can be independent, powerful allies to officials facing internal resistance to change. Members of Congress who often hold their seats much longer than senior military or political officials in the Pentagon can provide long-term support for specific programs that can help weather the changes in personalities in the DoD.

Congressional support to Admiral Hyman Rickover and the nuclear navy comes to mind as an early example. Today's F-16 and F18, programs that have together produced thousands of aircraft, were born out of the lightweight fighter prototyping program of the early 1970s. That program was forged by an alliance between a group of US Air Force officers, who were unhappy with their service's preference for large fighter aircraft that relied on increasingly sophisticated missiles to down enemy aircraft, and members on both sides of the Hill, who were interested in the ideas this group advanced as well as in using prototypes to test new ideas.[25] Much more recently, Secretary Gates was able to purchase the MRAP over objections from senior US Army officials by allying mainly with key individuals on the House Armed Service Committee.

On the organizational front, the Goldwater-Nichols Defense Reorganization Act of 1986 was the product of an alliance among members of the House and Senate Armed Services Committees and in some quarters in the Pentagon who were concerned about the lack of jointness in the US armed forces. Secretary of Defense Casper Weinberger showed no interest in reorganization, while Secretary of the Navy John Lehman actively opposed the act (and the navy showed no interest) as it evolved over the early 1980s. The Goldwater-Nichols Act would not have been passed without persistent and well-informed cross-Potomac collaboration. It remains controversial to this day, but no one doubts it was a sincere effort to induce more jointness in Pentagon thinking and training.[26]

There are hundreds of cases of this kind over many years, some very successful, many less so in terms of adding new and useful capabilities to the nation's forces. Each case is unique on several dimensions—the personalities of those involved, the broader status of defense on Capitol Hill, political alignments across Houses of Congress and with the White House, and the case for the technologies involved. But all have in common the power of an independent legislature, generally in alliance with Pentagon officials seeking change, to force change on services otherwise uninterested in the innovation in question.

Congress is charged with reviewing, marking up, and approving the defense budget. The cross-branch engagement with the Department of Defense, however, goes well beyond that, as members and staff on the

Hill and in the Pentagon interact on many levels to convey information and policy preferences and to push forward the policies they prefer. Ironically, what drives this level of engagement is the very independence of both branches. Without an overarching authority to make and enforce decisions, the only way to make things happen—or prevent things from happening—is to create cross-branch alliances that, hopefully, last over time.

CONCLUSION

Those unfamiliar with the defense budget process and watching it take place might marvel at a system of budgeting that brings so many considerations to bear on what might be expected to be straightforward strategic and technical decisions. Domestic political considerations, vastly differing views on defense policy, technical issues, and systems analyses all get rolled into a single budget. This is a complex business, and the marvel is that it has produced a remarkable force structure, sophisticated weapons, and an elaborate infrastructure, but at the same time a great deal of inefficiency and duplication.

Pentagon officials are likely to be less dazzled. To many of them, Congress is a nuisance; keeping members informed and keeping some of them happy are huge distractions from making analytically sound strategic policy. They would no doubt prefer to have more control over policy making and less interference. But that's not going to happen. Congress isn't going anywhere, and it certainly isn't going to legislate constraints on its exercise of constitutionally enabled oversight into the defense budget. Besides, Robert Gates's admonition that "we never once got it right" reminds us that no one—within or outside the Pentagon —has a monopoly on wisdom so far as US military strategy and force structure are concerned.

Pentagon officials have no choice but to deal with Congress, and the brief history outlined in this chapter suggests that they should do so with an eye for opportunities—the chance to shore up support, head off controversy if possible, and perhaps launch new initiatives. This makes managing defense a political undertaking far removed from the private sector world of bottom lines (short of war, at least) and executive power. But managing the DoD is not foreign territory; we have two centuries of experience with political management in the world the Founding Fathers created, over seventy years if one starts with the creation of the DoD in 1947. These tell us that, above all, political power in this system is a limited resource that can be spent, hoarded, or wasted. It does not grow on its own, and it is certainly not infinitely expandable. Defense managers have to maintain a very sharp sense of their priorities.

Robert Gates made this point in his memoirs when he noted that the Secretary of Defense "must be selective in identifying his [or her] agenda."[27] But the point had been made decades earlier in an article that belittled the effect of organizational changes in the DoD as far as the defense secretary is concerned:

> ...there is no internal organization, management system, or management philosophy that by itself will solve the problem of controlling defense resource allocation. The civilian leader's best hope for exerting influence over military capabilities is to make a selective and determined attempt to accomplish a few major goals in reforming weapons cost and performance where the economic, political, and military stakes are overriding.[28]

Harvey Sapolsky and his colleagues at MIT add the rather draconian point that "the rest of the vast department is left to run on its own with the hope that nothing too controversial or disastrous occurs."[29] But the fact is that managers elsewhere in the department are often doing the same thing.

Defense secretaries can operate on important margins, but short of war they probably can't tackle the big redundancies mentioned in the introduction to this chapter—two armies, five air forces, and so on. Perhaps the best that can be done here is to ensure that there is diversity in these forces. If the nation is going to have two or three versions of the same force, those forces should be doing different things, perhaps even competing with each other. Having two armies is wasteful, especially if they are both doing precisely the same thing.

NOTES

1. Kaufman, *The Administrative Behavior of Federal Bureau Chiefs,* 47.
2. Quoted in Wilson, *Bureaucracy,* 238 (emphasis added).
3. One is reminded of Harry Truman's comment about President-elect Dwight Eisenhower: "He'll sit here, and he'll say, 'Do this! Do that!' *And nothing will happen.* Poor Ike—it won't be a bit like the Army. He'll find it very frustrating." Quoted in Neustadt, *Presidential Power,* 9. While a fair assessment of the limits of presidential power, the quote reflects Truman's serious underestimation of Eisenhower's political acumen.
4. Robert Gates, Address to the West Point Military Corps of Cadets, Feb 25, 2011, https://www.americanrhetoric.com/speeches/robertgateswestpointspeech.htm.
5. U.S. Department of Defense, Secretary of Defense Speech, U.S. Military Academy, February 25, 2011. https://www.americanrhetoric.com/speeches/robertgateswestpointspeech.htm
6. The separation of *authorization* for spending from the actual *appropriation* of money in US practice dates from the early nineteenth century and has its roots in the eighteenth-century British parliamentary practice of separating policy from the actual distribution of money. See Schick, *The Federal Budget,* 191. Despite this long history, it is carried out unevenly in US practice in two ways. First, while defense spending has separate committees (armed services and appropriations subcommittees on defense), the same committee authorizes and appropriates some budget sectors (e.g., agriculture). Second, while technically "appropriated funds are to be spent according to the terms set in authorizing law," in practice authorizers have been known to appropriate, and appropriators have been known to authorize or simply ignore authorizations law. Ibid., 203.
7. "[R]egular appropriations were enacted after October 1 in all but four fiscal years between FY1977 and FY2019. Consequently, CRs have been needed in almost all of these years to prevent one or more funding gaps from occurring." U.S. Library of Congress, Congressional Research Service, *Continuing Resolutions* by McClanahan et al.
8. For a brief history, see Vincent Davis's chapter, "[The Evolution of Central U.S. Defense Management," in *Reorganizing America's Defense,* edited by Robert J. Art, Vincent Davis, and Samuel P. Huntington, 149–168.

9. The classic story of McNamara's approach to budgeting and his whiz kids was written by two of them, Alain C. Enthoven and K. Wayne Smith, in *How Much Is Enough.* (The RAND Corporation reprinted the volume in its original format in 2005). This volume remains, for many, the bible for today's systems analysts.

10. In fairness, many of these bases were hangovers from World War II and thus obviously "excess to need." Nevertheless, by 1977, Congress passed legislation effectively halting the Defense Secretary's ability to close bases without congressional approval. See U.S. Library of Congress, Congressional Research Service, *Base Closure and BRAC: Background and Issues for Congress*, April 25, 2019, p. 1, https://fas.org/sgp/crs/natsec/R45705.pdf.

11. Ibid., 109, 170–172, 263.

12. By 1965, Enthoven and Smith note that "the DPMs [Draft Presidential Memorandums, developed by OSD] had become the principle decision-making documents in the Defense Department." *How Much Is Enough?*, 55.

13. "Melvin R. Laird: Richard Nixon Administration," Historical Office, Office of the Secretary of Defense, https://history.defense.gov/Multimedia/Biographies/Article-View/Article/571291/melvin-r-laird/.

14. Even the name "systems analysis" became controversial. Thus, in the mid-1970s the name was changed to the Office of Program Analysis and Evaluation (PA&E), which had alongside it the "Cost Analysis Improvement Group." In 2009 the two offices were combined to make CAPE.

15. Cooper and Rumbaugh, "Real Acquisition Reform"; see especially the graph on p. 62.

16. Davis, *Framing the Problem of PPBS*, 5.

17. "[The services] not only have the power of the first draft; but are also the recipients of the funds once they are actually appropriated by Congress. Although the program and budget review by OSD lasts as long as six months, and the Congressional appropriations process takes nearly a year, recent experience indicates that over 90% of what the services request survives both processes intact." Ibid., 62.

18. Gates, *Duty*, chapter 4.

19. Russell Rumbaugh, "What We Bought: Defense Procurement from 2001 to 2010," Henry L. Stimson Center, October 2011: "Since FY01, we spent roughly $1 trillion on defense procurement, and the military services used that funding, including that provided in the supplemental war funding, to modernize their forces." (p. 16).

20. Just what motivates congressional votes on defense is hard to pin down. Defense contracts are assumed to be important, but the DoD does not keep data on defense contracts by congressional district, only by state, which eliminates the possibility for regression analyses of House votes. Overall, the literature suggests that while contracts are important to members, they come and go—lose one, get another. In contrast, installations—posts, air, and naval bases, depots, and so on—do not go anywhere and are often large and important communities within states and districts. They are thus extremely important. See especially Mayer, *The Political Economy of Defense Contracting*.

21. As Heidi Demarest points out in her book on the US Army and Congress, "most staffers and long-serving members of Congress are steeped in institutional knowledge of a program and are able to compare the current set of products with past information." As one Department of the Army Systems Coordinator marveled about this in mild disbelief, noting "She [a congressional professional staff member] had every briefing on my program for the last ten years. I don't even know where that stuff is." (*US Defense Budget Outcomes*, 158).

22. "The Gold Watch Syndrome" refers to a game the military services can use to seek additional funding of their budget. Service officials may remove a very popular program - a gold watch -- from their budget submission, calculating that members of Congress will reinsert it, hopefully providing additional money to fund the program rather than cancelling another program to return the service's budget to its original level.

23. With the Cold War over, Congress even devised a way—the BRAC process—to close bases, which are normally prize possessions in any state or district. But as successive BRACs have reduced the number of remaining "targetable" bases, congressional enthusiasm has waned despite widespread agreement that the nation's military infrastructure is still larger than required for today's forces.

24. Another game is buy-in/get-well; that is, "buy in" with especially rosy initial cost estimates and early-year spending, and then "get well" with rising annual budget requests after it is clear that Congress is committed to the program. Like the gold watch syndrome, this is difficult to pin down. Are initial cost estimates intentionally understated, or do they merely reflect the kind of optimism that even homeowners confront with something far less complicated than a weapon project?

25. Air force officials worried that an alternative fighter program would threaten the F-15 program just then going into production. Later, how-

ever, with F-15 production well underway, the service took Secretary Melvin Laird's offer to increase force structure and purchased the F-16.

26. For a short, totally obscure but reasonably accurate history, see Thomas L. McNaugher and Roger L. Sperry, "Improving Military Coordination: The Goldwater-Nichols Reorganization of the Department of Defense," in *Who Makes Public Policy? The Struggle for Control between Congress and the Executive*, edited by Robert S. Gilmour and Alexis A. Halley (Chatham, NJ: Chatham House Publishers, 1994), 219–258.

27. Gates, *Duty*, 577–578.

28. Lynn and Smith, "Can the Secretary of Defense Make a Difference?," 69.

29. Sapolsky, Gholz, and Talmadge, *US Defense Politics*, 59.

CHAPTER 8

STRATEGIC CHOICES IN DEFENSE FORCE STRUCTURE

Michael Linick

The balance we strike between capability, capacity, and readiness
will determine the composition and the size of the force for years
to come.

—Chuck Hagel, former Secretary of Defense[1]

INTRODUCTION

When planning specific military operations, either for actual use or as a
contingency plan, planners across the Department of Defense have the
relatively easy task of planning how to use fairly well-defined resources,
in commonly understood ways, to achieve fairly well-defined objectives
against a fairly definable foe in a known geographic location. The question

of when they will have to do this may be open, but the basic framework of what will be done and why is relatively static.

In contrast, when planning and resourcing national strategy to meet a wide range of strategic security demands, planners deal with tremendous uncertainty. Which potential enemy will become the active foe? When? What will be the war aims? Can they be deterred? What about black swan events? Given that, a well-written strategy must establish strategic ends and risk tolerance; it must also provide guidance on ways and means these ends will be accomplished. Unfortunately, this bridging of ends, ways, and means is often missing or poorly articulated in DoD strategies. To be effective, defense guidance should address four major topic areas, which constitute the four pillars of defense planning. They are:

Force Size and Composition
This includes both the overall size of the force and the mix of units/ capabilities that comprise it. (e.g., aircraft carriers, bombers, and Armored Brigade Combat Teams [ABCT].)

Force Posture
Posture is the mix of forces stationed overseas compared to those stationed in the United States. Posture includes forces rotating overseas rather than being stationed there. Understanding the force posture provides insight about how the strategy intends to use the military, both to prevent crises and to respond to them. From a resourcing standpoint, posture is also a critical variable as forces stationed overseas tend to cost more than those stationed in the US, while forces rotating abroad are the most expensive of all.[2]

Force Modernization
Modernization is about new technologies, concepts, and doctrines. How will they be researched, developed, and introduced? How will they be

fielded? Which units will receive new capabilities, and over what period of time?

Force Readiness

Readiness is the ability to provide capabilities required by the combatant commanders to execute their assigned missions.[3] This chapter is centered on understanding readiness in terms of how much and which parts of the force must be at peak performance at any given time, and on how much of the remainder of the force can be built to an acceptable level of readiness once a conflict or crisis has begun. The next chapter of this work will fully deconstruct the multi-faceted idea of defense readiness.

The core challenge of strategic planning is that each of these four pillars requires resources, which is a combination of both funding and bureaucratic/organizational focus and capacity. Therefore, for a strategy to be effective, funding and prioritization of each of these four factors must be integrated into a plan that not only clearly understands the trade space among them and articulates the acceptable risk tolerance in each area but also is executable within the DoD's budget.

Put another way, the US defense strategy must balance resources between competing—and not fully known—needs: it has to balance near-term demands for ready forces with long-term demands for modern forces; it must also balance the costs of stationing or rotating forces overseas with the potential savings of stationing them in the United States; and, finally, it needs to balance the cost of additional force structure with the associated costs of having to man, equip, and train that structure as well as continuing to modernize it for the future.

However, the strategic planning environment and processes in use by the Department of Defense are not optimized to feed information into this balancing act. Most DoD processes start with an assumption about one or more of the four pillars and do not vary those assumptions in their analysis of the others. As a result, there is relatively little understanding of how to trade among the pillars, and experts in one area have little understanding or awareness of how their focus area interacts with the others. This chapter illuminates the logic of these trades, their opportunity costs, and the risks associated with making these trades as well as the complexity and the interconnectedness of these choices.

THE FOUR PILLARS AND STRATEGIC CHOICES IN DEFENSE FORCE STRUCTURE

The concept of the four pillars is not formally codified in DoD thinking; they are a planning construct developed to support the writing of the 2018 National Defense Strategy (NDS).[4] Three of the pillars—size, modernization, and readiness—are generally referred to as the "iron triangle" inside the Pentagon.[5] However, posture is not included within the triangle. Nonetheless, where units are based has great impact upon the DoD's ability to deter adversaries and deploy rapidly to a conflict.

To test their proposed strategy, the 2018 NDS writing team wanted a tool to help them examine as many combinations as possible of potential future environments and defense strategies. RAND developed that tool into a game called "Hedgemony," which abstracts the thousands of decisions that the DoD has to make into the four aforementioned pillars.[6] The idea behind the game is that the four pillars represent the key "intermediate goods" that the Office of the Secretary of Defense (OSD) uses and in combination provide a "final good" of national defense or national security.

Most of the discussions and decisions in the departments and services are about the contributing goods and services that, in combination, produce these four key intermediate goods. Although this is an inexact description of what the OSD does, the abstraction of the four pillars proved a very useful way to focus thinking on defense strategy and the required resource trades.[7]

The framework of the four pillars rests on the idea that within the DoD, there are literally hundreds of "products" with no "cost" or "revenue" feedback stream to help identify clear winners and losers. That said, there may be a set of observed combat outcomes or operational simulations/analysis that do so. At the platform/system or capability level, analysis goes into which to develop, which to continue, and which to phase out. Given that the DoD is both the producer and the consumer of the capability, other information flows normally found in a "market" get muddied. The critical distinction is that capabilities such as B-1 bomber squadrons are not a "final good"; they aren't built simply to exist. They are built so that they can accomplish a mission in support of a strategic objective or outcome. Thus, the desired outcome is the "final product."

Concepts like "military effectiveness,"[8] or "deterrence" are close to a DoD "final good" or "final product." However, they are difficult to measure. It is also hard to determine the mix of capabilities that will provide those products—especially given the wide range of conditions under which they may have to be provided and the wide range of capability mixes that can be generated. Because of this, a useful proxy is a list of DoD "intermediate goods" that focus on force size, readiness, posture, and modernization. Employed in the correct combination they allow for the final product—using military force to achieve political goals.

Military planners begin by asking what needs to be done and then looking at multiple options to accomplish these goals. In doing so, they make high-level trades. If responsiveness is a key requirement, they can choose between options that have forces forward—referring to those positioned near contingency locations—and those that have forces at a higher state of readiness or easier to deploy (or some combination of the two). If military overmatch against the enemy is difficult to achieve, planners may choose to rely on a more modern or larger force. If overmatch is less difficult, planners may be satisfied with a less modern capability. If numbers matter most, planners might opt for a larger force —but for the services to pay for that larger force, they might have to delay acquisition and modernization and/or reduce some readiness levels.

For a variety of reasons, the DoD will never be able to produce a force the planners want: one that is overwhelmingly large, at the highest level of modernization across all capabilities, one where all units are ready and immediately available, or one that has exactly the right units already present in all needed places. Resources are limited. Further, the trades required to produce these different choices are not among service roles or individual capabilities; they are trades between overall size, posture, modernization, and readiness. It is these four "products" which national security planners use to provide the finished good of military effectiveness which translates into security, deterrence, and the ability to win wars.

DoD senior leaders often do discuss their strategies in the context of these trades. In 2013, during the Strategic Choices Management Review (SCMR), the DoD examined the impact of significant budget cuts. It was clear that the budget would shrink, but it was not clear by how much; estimates ranged between $150B and $500B over the period 2014–2019.[9] The DoD's analysis focused on trades across force structure and modernization as well as looking for savings in DoD overhead costs. In announcing the SCMR results, Secretary of Defense Chuck Hagel noted:

> The basic tradeoff is between capacity—measured in the number of Army brigades, Navy ships, Air Force squadrons, and Marine battalions—and capability—our ability to modernize weapons systems and to maintain our military's technological edge.[10]

He added:

> In the first approach, we would trade away size for high-end capability.... We would protect investments to counter anti-access and area-denial threats... and we would continue to make cyber capabilities and special operations forces a high priority. This strategic choice would result in a force that would be technologically dominant but would be much smaller and able to go fewer places and do fewer things, especially if crisis occurred at the same time in different regions of the world. The second approach would trade away high-end capability for size. The DoD would

look to sustain its capacity for regional power projection and presence by making more limited cuts to ground forces, ships, and aircraft. But it would cancel or curtail many modernization programs, slow the growth of cyber enhancements and reduce special operations forces. Spending cuts on this scale would, in effect, be a decade-long modernization holiday. The military could find its equipment and weapons systems—many of which are already near the end of their service lives—less effective against more technologically advanced adversaries. It would also have to consider how massive cuts to procurement and research and development funding would impact the viability of America's private-sector industrial base. These two approaches illustrate the difficult tradeoffs and strategic choices that would face the Department in a scenario where sequester-level cuts continue.... *The balance we strike between capability, capacity, and readiness will determine the composition and the size of the force for years to come.*"[11]

THE FOUR PILLARS IN DETAIL

Size and Composition of the Force

The size of the force can be directly measured by the number of uniformed military paychecks Congress authorizes the DoD to issue. In the FY2020 National Defense Authorization Act (NDAA), this included 1,339,500 authorizations for active end strength and 807,800 for the reserve components.[12] These figures do not include the civilian or contract work force —for which no real cap on authorizations is provided—because these positions are paid for within other appropriation categories.

The military workforce is expensive. Looking only at pay and benefits, it is expected to consume between 22% and 25% of the DoD budget between 2022 and 2024.[13] Additions to force structure also require procurement of new equipment—adding to the procurement budget (but not, necessarily the research budget) and to the operations and maintenance budget. All units require resources to exist and train. New force structure may also create demands for military construction funds.

Aggregate force structure numbers are ultimately an unproductive lens through which to evaluate strategy because these raw numbers don't equate directly to military capability. As a result, planners prefer to look at the numbers of specific capabilities provided by the aggregate structure. Traditionally force size is evaluated by looking at specific elements within each service. For example, the Heritage Foundation's Index of US Military Strength uses the numbers of army brigade combat teams (BCTs), marine battalions, navy ships in its combat fleet and air force aircraft (grouped by fighters and bombers).[14]

There are also recognized challenges with simple counts of force structure elements or platform numbers when assessing relative military power or strategic advantage.[15] This is because there are two different sets of balance within the overall set of force structure issues. The first centers around the absolute size of the force, including such factors as the aggregate number of divisions, wings, or carrier battle groups. The second centers around the composition of forces within that aggregate size, the proportion of specific aircraft, ships, or other types of equipment within the overall structure.

The DoD has a vested interest in shaping what capabilities the military services buy as well as the mix of capabilities provided between and across the services. The DoD has an equally important role in setting overall aggregate personnel levels and their distribution among the military services and the DoD agencies and activities. Reducing overall force structure is a common way to pay for new investments in modernization or to pay for cuts in the DoD budget. Conversely, sacrificing modernization to pay for end strength has also occurred, as has building force structure whose readiness could not be fully funded.[16] These decisions are sometimes initiated by the OSD and imposed on the military services, while at other times they have been proposed by the services themselves. The discussion of trade space will be further examined next.

Force Posture

Force posture is a reference to the physical location of military forces, the infrastructure that sustains them, and the policies for their use in peacetime operations. The term encompasses activities and agreements associated with partner nations designed to make the use of US forces more effective and/or efficient. The DoD instruction on global defense posture states: "GDP [Global Defense Posture] processes apply to DoD's forces, footprints and agreements that support joint and combined global operations and plans in foreign countries and US territories and in defense of the homeland."[17]

An alternative definition of posture is based on presence and persistence:

> Presence is "[the] effect, or the act of being materially present in a specific geographic place at a particular moment in time," [... and] persistence is how long one can stay, combined with the influence created. Influence includes awareness, knowledge and reach.[18]

These definitions are useful in understanding the relative merits of land-based presence (including land-based airpower) and sea-based presence. Each requires a different mix of infrastructure to support, and each provides different levels of effects, persistence, and influence.[19]

Basing and infrastructure can come in several different categories. All military units have a home base, either in CONUS (the continental United States) or overseas. Units can also be deployed from their home base and operate somewhere else. Deployments can be short term (a few days or weeks) or much longer (three or more months). Overseas facilities supporting posture might be equipment storage sites or supply and maintenance depots managed by the US military, but with most of the labor coming from local nationals. There also may be agreements for shared use of military and/or logistics facilities.[20]

Policies that affect posture include the DoD's Deploy-to-Dwell/Mobilization-to-Dwell policies (how long forces can be deployed overseas)[21] and medical, training, or equipping requirements. These all affect when or if a servicemember or unit can be deployed. Further US mobilization authorities and agreements between host countries and the US are substantial variables. These affect the costs as well as interactions between US forces and host nation or other regional forces and impose limitations on the operations that US forces can execute.

The decision whether to station forces overseas (or to pre-position equipment or supplies there), to have them forward in a rotational mode, or to keep them at home and deploy only as needed is tied to several different political, warfighting, and cost criteria. The benefits of forward posture are clear if we can correctly identify where the force will need to be in time of crisis. RAND Corporation reports that "overseas presence contributes to contingency responsiveness, deterrence, assurance, and security cooperation."[22] The specific capability type chosen for presence may also matter. For example, BCTs, marine battalions, and permanently stationed aircraft provide different effects than carrier battle groups. Similarly, occasional rotations of ground forces provide different effects than do permanently stationed ones.

The financial costs of using rotational presence are generally believed to be even higher than the costs of overseas basing.[23] If force structure costs are included, then forward presence is also very expensive.[24] However, calculating the specific costs of posture out of the DoD budget submissions is exceedingly difficult. Most of the costs are buried in Operations and Maintenance (O&M) accounts which pay for things like overseas base operations. However, service and joint exercises with partner nations also fall under posture, as do the annual Overseas Contingency Operations (OCO).[25] These are all in different budget lines. Additionally, the cost of military inter-theater lift (air and sea) is a cost of posture, and this includes costs to pay for the development, procurement, maintenance, manning, and employment of the strategic air and sea lift forces.

Thus, the costs and benefits of posture (and how they vary between the specific force elements used to create that posture) are difficult to measure and are affected by other strategic considerations. RAND notes that "there is a minimum threshold of foreign posture that the United States must retain. Beyond that, there is additional posture that is almost certainly advisable to retain or even add... there are a number of choices specific to each region, where different judgments could lead to differing calculations of the advisability of reductions, additions, or changes in the nature of posture."[26] Among those calculations are considerations of the readiness, size, composition, and modernization of the force.

Modernization

Typically, military modernization refers to the replacement of an older technology with an improved version or to the introduction of a new technology with a military application. However, force modernization is more than just inserts of new or better technology. The 2018 National Defense Strategy explains:

> Modernization is not defined solely by hardware; it requires change in the ways we organize and employ forces. We must anticipate the implications of new technologies on the battlefield, rigorously define the military problems anticipated in future conflict and foster a culture of experimentation and calculated risk-taking.[27]

Modernization, then, includes a wide range of activities that can generally fall under the acronym "DOTMLPF:" doctrine, organization, training, materiel, leadership, education, personnel, and facilities. Each of these factors contributes to the overall employment of a capability. One can look to the differing tank doctrines and designs generated by Germany, France, Britain, and the United States between World War I and World War II as an illustration of why these variables matter. Each country had essentially the same technology (tanks, radios, airplanes). Each combined them differently, with wildly different results.[28]

Joint Publication 1-02—The Department of Defense Dictionary of Military and Associated Terms—defines doctrine as "fundamental principles by which the military forces or elements thereof guide their actions in support of national objectives. It is authoritative but requires judgment in application."[29] The Army War College provides the following definitions for the remaining elements:

- Organization: how we build structures of people and equipment to fight....
- Training: how we prepare to fight tactically...
- Materiel: all the "stuff" necessary to equip our forces so they can operate effectively...
- Leadership: influencing people by providing purpose, direction, and motivation, while operating to accomplish the mission and improve the organization.
- Education: how we prepare our leaders to lead the fight.
- Personnel: those individuals required in either a military or civilian capacity to accomplish the assigned mission.
- Facilities: real property; installations and industrial plants that support our forces.[30]

In DOTmLPF, the "m" or "materiel" piece is often written in lowercase. This is meant to be a reminder that improving materiel—equipment—is the most difficult, time-consuming, and costly option for modernizing. Changes in any of the other DOTMLPF dimensions are less costly to implement. However, there are clearly times when technology is the only answer. Doctrinal, training, or leadership changes can also take some time to create the desired effects on force capability. What is important to understand is that modernization takes time and that the amount of time it takes is uncertain, as is the final effect it creates.

As a pillar of defense planning, the modernization of systems and technologies has been prioritized. From 2014 to 2024, modernization consumed (or is budgeted to consume) between 29% and 35% of the DoD budget.[31] Within that budget, the DoD is both looking to expand the capability range in which technology can deliver (i.e., military applications for existing technology) as well as find, develop, design, and test specific applications of technology that can provide military utility.[32] Balancing this research between nascent, emerging, and current technologies, and then deciding which technologies to focus on are key issue areas in the DoD's and the services' strategy discussions. Decisions made affect the speed with which more modern capabilities are introduced, both to the first unit equipped and then to the broader force. They also affect the level of leap ahead that any new increment of modernization provides.[33]

Once a technology has been developed in laboratories, it must then be manufactured and distributed. Every year the DoD procures a combination of the newest technology and older systems. For example, Army AH-64E procurement (the most modern version of the Apache helicopter), is scheduled over a ten-year time frame and includes both new production and conversion of older helicopters.[34] Another example is Virginia-Class submarines which are being procured over a twenty-year-plus time horizon, although the design continues to be modified and modernized even while the procurement continues—older models will be retrofitted with the newly developed and integrated technologies.[35]

In summary, technological change consumes about one-third of the DoD budget. However, modernization is more than just about changing the technology in the force. The rate of technological change and the effectiveness it will provide are both uncertain and entail risk—but there is also a potentially high payoff. In fact, the DoD budget for new technologies is, in many ways, a classic venture-capital hedging strategy —it invests in everything from new science to new applications of proven science and the development and refinement of existing systems. The assumption is that enough of them will be sufficiently successful, in a relevant period, to provide technological parity and, ideally, technological superiority against US adversaries.

Modernization impacts force size and composition not only because it competes for a share of the budget but also because more expensive systems cannot be procured on a one-for-one basis with the systems they replace. Therefore, buying advanced fighters or submarines often means that over time the force has fewer of each (albeit the ones they have are more capable). Modernization can also affect readiness as modern equipment requires more maintenance than what it replaced, and units require training in greater volume or rigor to achieve proficiency with their new equipment. Modernization can impact posture decisions, such as sending a political statement by forward stationing the most modern equipment or modernizing equipment sets stored forward.

Readiness

Readiness is a difficult term of art in the DoD that is covered in much greater depth in chapter 9 of this volume. One could argue that everything the department pays for contributes to readiness (generally, every member of the DoD who is arguing for budget share will likely aim to get the word "readiness" into their budget justification). This chapter focuses on the immediate operational readiness of military units to execute their mission. This type of readiness has two components—the missions assigned to a unit and the amount of training the unit has had on those missions. If there is a mismatch between what units have been training to do and what they are being asked to do, then it is likely that those units will need additional time before they can be deployed for the new mission. Adding to these considerations is that training decays, both because muscle memory fades when the muscles are not being put to use and because the members of the teams change over time and the new members need to train together.

It is also important to consider the distribution of units maintained at a high level of readiness—those whose mission is to literally "fight tonight"—versus those that will require some time to prepare, either because of overall readiness or because of the need to retrain on a different mission set. The bottom line for resourcing is that a more responsive and ready force is more expensive than one that is relatively less so.

It is not really possible to maintain all units at high levels of readiness. Each of the services have different ways of rotating readiness through units, so all of them have an inventory of more and of less ready units. People move to new jobs or go to required military schools, and when they do so, unit readiness falls. Equipment may need to go to depots or shipyards for more extensive repair and refit, making units unavailable or at least less ready. Personnel need a break from high operations tempos to reset, or military retention rates will fall. Units also require time to train on new equipment and/or new tactical concepts. Additionally, a large part of the military force is in the reserve component, and, almost by definition, these units are less ready than their active component counterparts simply because these units have less time available to train.[36]

The services and the DoD have to think carefully about how ready is ready enough. What are the likely needs for immediately ready forces (by unit or platform type) and where can they accept risk, assuming there will be enough time available to bring units to a higher-than-anticipated state of readiness? These decisions can be impacted by posture (how long it will take to get the unit to where it is needed), modernization (how capable does the unit need to be, given the specific mission it is being given), and force size and composition (how many like units are there, and how many of them are ready now).

Trade Space

The preceding sections presented the four pillars separately. This section focuses on how to think about the necessary resource trades among those pillars. In *Military Readiness,* Richard Betts posits that strategic planners need to be able answer the three questions of: ready for what, ready with what, and ready for when.[37] Betts's exposition is essentially about the trade space between time and capability. If we need it now, we prioritize forces that are immediately ready. If there is time for a build-up (e.g., Desert Storm) then readiness can be built as needed (which is the central argument for a reserve component). If readiness is not immediately required, there are more efficient ways to fund national defense—allowing modernization to build efficiency and temporarily reducing force structure to create savings.

But, as discussed earlier, how quickly something can be brought to the conflict also depends on where it starts. Thus, posture is a key factor in responsiveness. So, if we know where the fight will be, and we can build the alliances, forward station the troops, and maintain their readiness; then forces will be much more responsive—even forces that are less ready. Conversely, if it will take three months to get a meaningful force forward, then an enemy's calculation will entail whether they can achieve their aims in under three months, or at least have raised the cost of evicting them to be higher than the US is willing to pay.[38] Additional investment in strategic lift may help shorten the deployment timeline, but, to paraphrase Stalin, presence has a quality all its own.[39]

Of course, being ready in one potential geographic theater of war does not guarantee readiness in another. So, to be forward everywhere requires a larger force. Finding the right mix between where and with how much requires risk-informed decisions and an assessment of how quickly relevant forces can be moved. The relevance of the forces is critical. Air power can generally deploy most quickly, sea power can generally deploy relatively quickly, and even light ground forces—such as the 82nd Airborne or a Marine Embarked Unit (MEU) may move quickly —but all may have significant operational challenges.[40]

If we don't know when the fight will occur, we will have to be concerned about both the enemy's ability to continue to improve their capability and our own ability to free up funds for modernization—to preserve or regain overmatch. Returning to Stalin's thinking: quantity does have a quality all its own, and quantity can, in some cases, make up for shortages in capability. However, enough of a quality differential can overrule even large quantitative deficiencies at some point.

A second set of trade decisions arises when we are not sure of when or where the fight will be. The concern then becomes how to balance a modernization portfolio between capabilities and between near-term and long-term investments. Delays in modernization in one area may require risk mitigation in the form of additional size, different posture, or enhanced readiness. For example, in the Baltics, NATO may not believe they can develop the new capabilities to defeat Russian anti-access/ area denial (A2AD) capabilities along the Russian border areas in the near term. So, NATO in the interim may decide on greater presence and readiness until counter A2AD capabilities have been developed. And, if the costs of this near-term additional presence and readiness are acceptable, that may imply that modernization dollars (for now) can be used to improve some other needed capabilities.

CONCLUSION

The four pillars of strategic planning discussed are not offered as a substitute to a nuanced understanding of the multiple entrenched and intertwined process that constitute defense strategic planning. Instead, they are meant to suggest a framework within which national security professionals can have a more effective discussion of strategy. DoD experts often work inside programmatic stovepipes. Thus, these experts can say, with some level of certainty, how a change to the budget for spare parts will affect readiness or how moving a squadron from Germany to the United States will affect deployment timelines in the event of a Baltic contingency. They also know where the DoD does or does not have technological overmatch or how much it will cost to grow the force by 10,000 spaces.[41] But these same experts are not able to explain—and often are not even asked to think about—how their recommendation in one program area (e.g., modernization) may affect another (e.g., posture).

There are officers both in the OSD and the military services that do have the responsibility to integrate across these trades, but outside of a major effort like the SCMR, those discussions are often perfunctory, and the analysis is not rich. DoD budgets have a certain path dependency. On an annual basis, they can only be changed at the margin, although admittedly, in some years the margin is significantly larger than in others. As such, the focus tends to be on the marginal changes at the program level.

Teaching analysts and strategists to think in terms of the four pillars is critical to linking the ends, ways, and means of defense strategic planning. Requiring budget proposals and strategies to expressly discuss priorities in terms of size, posture, readiness, and modernization as well as the trades between them can only help elevate the quality of understanding of what a budget does do and what the strategy demands—and whether the two are in balance.

Notes

1. *Statement on Strategic Choices and Management Review*, as delivered by Secretary of Defense Chuck Hagel, Pentagon Press Briefing Room, Wednesday, July 31, 2013, https://www.defesanet.com.br/geopolitica/noticia/11665/US-DOD---Statement-on-Strategic-Choices-and-Management-Review/.
2. Lostumbo et al., *Overseas Basing of U.S. Military Forces*.
3. JP 1-02, DOD Dictionary of Military and Associated Terms, as of June 2020, https://www.jcs.mil/Portals/36/Documents/Doctrine/pubs/dictionary.pdf.
4. This statement is my personal recollection from being involved with the writing process of the 2018 National Defense Strategy.
5. Frank Hoffman, "Scraping Rust from the Iron Triangle: Why the Pentagon Should Invest in Capability," *War on the Rocks*, February 9, 2018, https://warontherocks.com/2018/02/scraping-rust-iron-triangle/.
6. Information about Hedgemony (and, in fact, the rules of the game) can be found at https://www.rand.org/pubs/tools/TL301.html. I was the lead designer for the game and assisted in running it for OSD's Policy directorate as they explored options for the 2018 NDS.
7. The basic concept behind this abstraction was influenced by RAND economists advising the Hedgemony team, who introduced the concept of activity-based costing as a way to think about the problem set.
8. Stephen Biddle notes that "Military effectiveness is defined as the ability to produce favorable military outcomes per se, including the outcomes of minor skirmishes at the tactical level of war and the outcomes of wars or even long-term politico-military competitions at the strategic or grand strategic levels of war. See "Military Effectiveness," *Oxford Research Encyclopedia of International Studies*, https://doi.org/10.1093/acrefore/9780190846626.013.35.
9. James Carafano et al., "Pentagon Strategic Choices and Management Review: Early Warning Two Years Too Late," The Heritage Foundation, Aug 2, 2013, https://www.heritage.org/defense/report/pentagon-strategic-choices-and-management-review-early-warning-two-years-too-late.
10. *Statement on Strategic Choices and Management Review*
11. Ibid. Emphasis added.

12. These numbers, broken out by component can be found in the annual National Defense Authorization Act (NDAA). For the FY20 numbers, see 116th Congress, *National Defense Authorization Act for Fiscal Year 2020*, January 3, 2019, Subtitle A, https://www.congress.gov/116/bills/s1790/BILLS-116s1790enr.pdf).

13. Using both the MILPERS and the Family Housing costs as the cost of military personnel.

14. "Introduction: An Assessment of U.S. Military Power," 2021 Index of U.S. Military Strength, The Heritage Foundation, November 2020, https://www.heritage.org/military-strength/assessment-us-military-power.

15. See, for example, The Heritage Foundation, 2021 Index of U.S. Military Strength, https://www.heritage.org/military; Andrew Marshall, "Problems of Estimating Military Power," https://www.rand.org/content/dam/rand/pubs/papers/2005/P3417.pdf; and Bastian Giegerich and Nick Childs, "Military Capability and International Status," IISS, https://www.iiss.org/blogs/military-balance/2018/07/military-capability-and-international-status.

16. For some comparative discussions about these kinds of trades, see, Larson, Orletsky, and Leuschner, *Defense Planning in a Decade of Change.*

17. U.S. Department of Defense, *Management of U.S. Global Defense Posture*, DoDI 3000.12, May 6, 2016, with Change 1, May 8, 2017, 4. DoDI 3000.12, May 6, 2016; Incorporating Change 1 on May 8, 2017 (whs.mil)

18. Hendrix and Armstrong, *The Presence Problem.*

19. See Deni, *The Future of American Landpower.*

20. The exact number of U.S. troops and bases abroad varies year to year. However, the number remains substantial. As of March 2021, there were more than 170,000 US servicemembers stationed or operating overseas. For more information, see Jane Siebens et al., "U.S. Global Force Posture and U.S. Military Operations Short of War," The Stimson Center, July 14, 2021, https://www.stimson.org/2021/us-global-force-posture-and-us-military-operations-short-of-war/.

21. See Under Secretary of Defense, *Memorandum for Secretaries of the Military Departments, Chairman of the Joint Chiefs of Staff, Subject: Under Secretary of Defense (Personnel and Readiness) Deployment-to-Dwell, Mobilization-to-Dwell Revision*, OUSD, 1 Nov 2013.

22. Lostumbo et al, *Overseas Basing of U.S. Military Forces.*

23. Ibid., 131–165.

24. For most forward rotational presence, the need to maintain a rotational presence forward implies that there are at least two similar capabilities

either preparing to assume the next rotation or recovering from the last one. This additive requirement for both readiness expenditure and force structure can be quite costly. See, for example, Deni, *Rotational Deployments vs. Forward Stationing.*

25. Over the five years FY2015–20, the range of the OCO budget was $58–$68 billion. (See Department of Defense Green Book, chapter 2, https://comptroller.defense.gov/Portals/45/Documents/defbudget/FY2022/FY22_Green_Book.pdf).

26. Lostumbo, et al., *Overseas Basing of U.S. Military Forces*, 303.

27. *Summary of the 2018 National Defense Strategy of the United States of America: Sharpening the American Military's Competitive Edge*, Department of Defense, Washington, DC, undated, https://DoD.defense.gov/Portals/1/Documents/pubs/2018-National-Defense-Strategy-Summary.pdf.

28. For a fascinating discussion about these different approaches, see, David E. Johnson, *Fast Tanks and Heavy Bombers: Innovation in the U.S. Army 1917–1945* (Ithaca, NY: Cornell University Press, 2003).

29. Joint Publication 1-02, *Department of Defense Dictionary of Military and Associated Terms*, November 8, 2010 (as amended through February 15, 2013), DoD, Washington DC, Feb, 2013, https://code7700.com/pdfs/jp1_02.pdf.

30. Chairman of the Joint Chiefs of Staff Instruction CJCSI 3170.01H Joint Capabilities Integration and Development System, January 10, 2012, uses the expanded acronym DOTmLPF-P, where the terminal letter "P" stands for "policy." For simplicity, I have used simply DOTMLPF.

31. Statement by Secretary of Defense Lloyd Austin on the President's Fiscal Year 2022 Defense Budget, May 18, 2021, https://www.defense.gov/Newsroom/Releases/Release/Article/2638711/the-department-of-defense-releases-the-presidents-fiscal-year-2022-defense-budg/.

32. For an interesting perspective on the origins of the Internet, see Barry M. Leiner et al., *Brief History of the Internet*, https://www.internetsociety.org/internet/history-internet/brief-history-internet/. It also provides a useful case history on how technologies mature from concept to application to iterative improvement.

33. "Leap-head technology" is a term by the DoD that is is generally used to refer to something entirely new that provides a tremendous advantage over any adversary.

34. DODIG Report-2018-130, *Procurement Quantities of the AH-64E Apache New Build and Remanufacture Helicopter Programs*, DoD IG, Washington DC, June 25, 2018.

35. Nathan Gain, *U.S. Navy Receives 1st Virginia-Class Block IV Nuclear-Powered Attack Submarine From GDEB, Naval News*, April 18, 2020, https://www.navalnews.com/naval-news/2020/04/u-s-navy-receives-1st-virginia-class-block-iv-nuclear-powered-attack-submarine-from-gdeb/ and U.S. Library of Congress, Congressional Research Service, *Navy Virginia (SSN-774) Class Attack Submarine Procurement: Background and Issues for Congress*, RL32418, (June 24 , 2019), https://assets.documentcloud.org/documents/6167505/Navy-Virginia-SSN-774-Class-Attack-Submarine.pdf.

36. As of June 30, 2021, the OSD reported 799,903 service members assigned to all reserve components. The Department of Defense produces a monthly unclassified report on personnel that is available at https://dwp.dmdc.osd.mil/dwp/app/DoD-data-reports/workforce-reports.

37. Betts, *Military Readiness*, 24.

38. Shlapak and Johnson, *Reinforcing Deterrence on NATO's Eastern Flank*.

39. There is apparently some question as to whether Stalin ever said, "Quantity has a quality all its own." See Mathias Klang, "Quantity has a quality all its own," http://klangable.com/blog/quantity-has-a-quality-all-its-own/, which attributes it to Thomas Callaghan.

40. See, for example, Schwarzkopf's discussion of ground options involving 4,000 soldiers of the 82nd Airborne Division in the early part of what became Desert Storm, "Their job would be to assert a U.S. presence—a dangerous mission, because if Iraq attacked, the 82nd Airborne's light weapons would be no match for Saddam's tanks. Within two weeks we could more than triple that ground force.... by the end of the first month our so-called "heavy" units would start to arrive...." H. Norman Schwarzkopf and Peter Petre, *It Doesn't Take a Hero*, 301.

41. The DoD authorizes force structure by spaces for required personnel, which are then in turn filled by accessions.

CHAPTER 9

RESOURCING
MILITARY READINESS

Laura Junor Pulzone

For the want of a nail the shoe was lost.
For the want of a shoe the horse was lost.
For the want of a horse the rider was lost.
For the want of a rider the message was lost.
For the want of a message the battle was lost.
For the want of a battle the kingdom was lost.
And all for the want of a horseshoe nail.

—13th-century German proverb[1]

INTRODUCTION

One of the toughest challenges for the Department of Defense is determining how to allocate its limited resources to counter the spectrum of

current and future threats. The preceding chapter explained the strategic choices that senior defense officials face in resourcing the iron triangle of programming, which is defined as force structure, modernization, and military readiness. This chapter extends that discussion with a deep dive into the concept of readiness, defined in the DoD dictionary as "the ability to provide capabilities required by the combatant commanders to execute their assigned missions...."[2]

Maintaining ready forces is a daunting challenge because the mechanisms that produce ready military capabilities are enormously complex; and the choices, some strategic (prioritizing one threat over another) and some tactical (prioritizing the remediation of one particular readiness deficiency over another) are rife with uncertainty and risk. As budgets shrink and the character of warfare continues to evolve, national security leaders need a specific accounting of the readiness limits of the force, the consequences of those limits and the insight to make timely and effective mitigation decisions. As threats evolve, the Department of Defense must ensure that the US military maintains its technical superiority.[3] That emphasizes the need to understand the intertemporal aspects of decision-making—the effect of today's choices on the likelihood that forces in ten or twenty years will be capable of prevailing against peer threats. The key to successful readiness management and the identification of constructive mitigation strategies in this environment is an informed debate about causes, effects, opportunity costs, and ultimately a clear, candid conversation about risk.

Geopolitical frames matter, and this is particularly true with regard to readiness. In his work *Military Readiness*, Betts asks three questions: "Readiness for When?," "Readiness for What?," and "Readiness of What?."[4] The geopolitical frame of the 2018 National Defense Strategy (NDS) is great power competition.[5] As such, the DoD is currently framing its required military readiness in these terms.[6]

THE CONCEPT

The analytic framework described here is straightforward and involves two basic steps: understand how the supply of ready forces is produced and then evaluate the available quantity of ready military capabilities supplied against the quantity of those forces demanded. The quantity supplied refers to how many ready military capabilities (from dog teams to long-range strike sorties to carrier battle group presence) the military services can produce. The demand refers to the planned and unplanned requirements for those capabilities. A readiness deficiency exists when the Department of Defense cannot generate the ready capabilities for these discrete requirements.

While the concept of this framework is simple, assessing the hundreds of production processes that underlie it is a challenge. That said, these production processes are well-understood ones that are already monitored and frequently quantified using detailed data from transactional databases. For most demand scenarios, the planners either have or can estimate the quantity of forces demanded throughout the course of a given operation. In other words, the infrastructure to understand these processes already exists. The Department of Defense has a surprising amount of data, some of which is extraordinarily detailed and cataloged and if arranged and assessed properly, can yield a lot of information.

In order to gain a clear picture of the problem, it is useful to map out the production processes. Consider the old proverb "for the want of a nail."[7] The idea here is that if people understand the process and dependencies in readiness production, they can protect the kingdom from being lost due to a seemingly innocuous problem like a missing horseshoe nail. Simply laying out the production flows highlights process vulnerabilities, critical dependencies, and throughput constraints, even without amplifying data and advanced statistical methods. For a position like a fighter pilot, the production process looks like convincing smart, healthy young people to choose the military as a career, then giving them years of education and highly technical training in various aircraft with different weapons systems and using varying training ranges and simulators until they are proficient and deployable.

We can also determine critical inputs. Continuing with this example, the production of ready pilots requires young people willing and able to serve in the military as pilots; working aircraft, simulators, and other systems for them to operate; and access to training opportunities. We know that protecting these critical inputs is essential to protecting pilot production. Conversely, any degradation in these inputs will (eventually) limit production.

Finally, we can determine how long the production process takes. In addition to considering how seemingly distant inputs might impact final production, understanding these production processes means also having to estimate how long it will take for the full effect to be realized. In the pilot example, counting from the day these young people join until the day they attain the desired proficiency, the process takes years; less for a new "nugget" pilot than for an advanced pilot. We can look at the status of the production line and critical inputs and get an idea of what production will look like in the near future.

With this simple example in mind, consider the wealth of information available for readiness management:

- The known number of pilots at each level of proficiency that DoD has now.
- The estimate of how many more will be produced over the next few years, assuming things stay constant (the ceteris paribus assumption in economics).
- A reasonably informed prediction of how problems with the production pipeline will affect the inventory of ready capabilities.
- The requirement to monitor this production pipeline closely and protect the critical inputs as fighter pilots are a critical military capability.

With this production information, one can compare the quantity of ready force supplied now and in the near future with the anticipated operational demand for those forces to determine a) if a supply shortfall exists, b) the operational risk associated with any given shortfall, and c) the driver of a given readiness deficiency. What follows describes an analytically based supply-demand analysis for each critical military capability that allows us to assess the sufficiency of military capabilities for various operational requirements.

The Supply of Ready Forces

It is impossible to manage readiness effectively without understanding the processes that produce military capabilities. The more detailed this understanding, the more opportunities there are to protect the processes, influence the rate of production, and forecast near-term production. The DoD generates ready forces (i.e., produces military capabilities) using a series of interconnected production processes, or pipelines, that combine labor (like mechanics and aspiring pilots), equipment (like airframes, spares, and ordnance), and training (including the time and space to train for various competencies) to produce a military capability (like a deployable long-range strike package). They typically begin with raw inputs (e.g., recruited personnel) that produce an intermediate good (e.g., a trained pilot or working aircraft), which serves as an input into making the final product (e.g., deployable long-range strike capability). This logic borrows straight from classical theory of production.

Because the complete processes are so complex, it is helpful to think about the production in terms of the continuum of interconnected production pipelines. Upstream portions of these pipelines range from raw materials to intermediate goods. In the pilot example, these are the processes that produce aviation maintenance personnel, the number of available simulators, and depot-repaired aircraft. The downstream portion of pipelines gets closer to the production of final products, such as deployable long-range strike capabilities for a specific operation.

Figure 1. Upstream and Downstream Portions of a Complete Production Line.

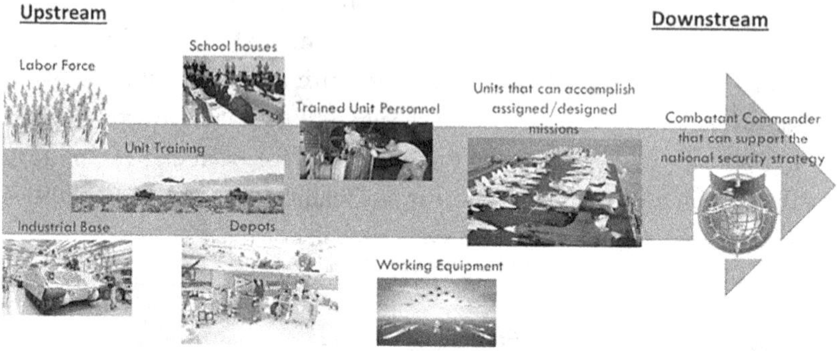

Source: Original Work by the Author

Figure 1 illustrates the upstream and downstream portions of a complete production line. Focus on the downstream end of the figure and consider that the ultimate downstream metric is whether the combatant commands can execute the operations expected of them in the current version of the National Defense Strategy.[8] The combatant commands rely heavily on the military services to provide the necessary capabilities like reconnaissance and strike. In many cases, military planners have estimates of how much of these capabilities would be required and when—these are elements of the demand signal that will be covered in the next section. For now, consider a particular demand signal that requires five long-distance strike sorties the first day of operations and another twenty more within the first week. The navy and air force both produce this capability and can monitor how many strike packages are trained and ready for that mission now, and how many more could be ready over an extended period. Working further upstream on figure 1 (backward or to the left), services can report the overall mission capability status of critical equipment, the availability of key ordnance, maintenance depot throughput, and even basic information on the inventories of critical personnel that are relevant to the production of ready strike packages.

Monitoring Readiness

There are a few categories of readiness metrics that managers should monitor for significant change. Each of these readiness elements represent nodes in the overall production pipeline that generates long distance strike packages. If one understands the inventories and production rate at each node, one has a comprehensive picture of the status of a particular military capability now and—if nothing else changes—in the near future. Readiness managers can use these metrics as vital signs to gauge a snapshot of the comprehensive readiness picture for distinct military capabilities.

Monitor Both Inventory and Production Rates at Each Pipeline Level

Managers ultimately care most about the downstream inventories since that is what the Department of Defense has to operate with. However, it is the level of intermediate goods that feed that downstream production that will provide the most valuable forecast information. For example, are the training schools getting enough high-quality students to feed potential pilots in training squadrons? Are the training ranges getting enough operable simulators and other enablers to fully train both individual pilots and complete strike packages? Are the depots repairing aircraft fast enough to keep squadrons flying? Using this example, monitoring production at each node will provide valuable information on the readiness of these strike packages because problems upstream will eventually cause problems downstream. Does one observe fewer working aircraft or reduced training opportunities? Such issues will certainly restrict the number of final strike capabilities that the navy or air force can produce. Depending on the production times involved, the effects could be seen relatively quickly (months) or over the course of years.

Monitor Critical Inputs/Drivers at Each Production Node

Some inputs/drivers will have more significant effects on the final product of a military capability than others. In some cases, downstream production is extremely sensitive to changes in a particular input or intermediate good. In other cases, the effect is difficult to mitigate. Consider the difference between having too few consumable spare parts and having too few trained pilots. The service could solve consumables shortages by purchasing more spare parts from the private sector. The effect on producing strike capabilities will be mitigated as soon as any resulting maintenance backlog is processed.

Pilots, in contrast, must be "grown" from within the service precluding an immediate solution. If managers are not directly monitoring the total inventories of junior and senior pilots as critical inputs and are instead only monitoring the mission readiness of deployable units, then it could take as much as a year before downstream managers take note of a reduced number of ready squadrons.

Upstream Processes Deserving Special Consideration

There are two classes of upstream processes that deserve special attention. One is the production of personnel that are both high in quality and sufficient in numbers, as used as an example in this chapter. The second is the collection of processes that produce working, mission-capable weapons systems.[9] In the aforementioned example that compares the effects of pilots and consumables, both are described as having the potential to impact downstream readiness. However, in the case of the pilots, the effect is likely to be more serious because the shortage of this input is difficult to mitigate. For that reason, inputs like inventories of critical, highly trained personnel should be carefully managed. Note that the quality of personnel, independent of the quantity, is itself a significant readiness driver.[10] Over time, units composed of high quality personnel—measured in terms of factors such as entry-exam scores and experience—are consistently associated with better readiness outcomes as measured in terms of lower equipment failure rates, faster repair rates, or better or timelier training.[11] This finding is especially relevant in assessing the value of lowering recruit quality (e.g., allowing for recruits with lower test scores) as a means of remediating shortages in the quantity of personnel.[12]

Problems with equipment maintenance can have a similarly insidious effect on the ability to generate a sufficient stock of working equipment to meet unit requirements. A unit's stock of working equipment is in turn determined in part by the maintenance system's ability to generate that working equipment. Any factor that limits the ability to either perform preventative or remedial maintenance on weapon systems will, holding other things constant, certainly create a backlog of repairs until units no longer have enough working equipment to adequately train and deploy. If one is not monitoring the entire production line, all that would be visible is a combination of units that are not suitably trained or units with poorly functioning equipment. The trigger, however, likely happened much earlier with something benign like a reduction in the number of maintainers, shortages in spare parts, or aging air frames.

DoD Readiness Metrics

The informed reader is likely wondering about the role of traditional readiness metrics like unit-level Status of Resources and Training System (SORTS) scores and Defense Readiness Reporting System (DRRS) mission and task assessments.[13] These are indeed useful for monitoring extreme downstream readiness like the status of deployable units (e.g., ships, battalions, and squadrons) and the ability of joint organizations, such as a combatant commander's ability to execute assigned missions. In fact, for monitoring downstream readiness, the DRRS is a critical metric that joint commanders use to assess whether they can conduct their assigned missions. SORTS and DRRS assessments, however, will never be sufficient management tools on their own because by only focusing on the downstream elements of the production processes, they can only be lagging indicators of force generation or readiness capabilities. In other words, they are only useful in identifying problems that have likely been existent for months or years and are just now manifesting because of the inability of a unit to perform an assigned mission.

Diagnosing Problems and Identifying Mitigations.
Another critical management tool is diagnostics and the development of mitigations. Figure 2 is a detailed depiction of a system of four production processes that illustrates how to trace the effect of a problem in recruiting/ training maintenance personnel for the generation of strike sorties. The trace can also work the other way. If, despite the best monitoring efforts, we failed to forecast a decline in downstream readiness, this framework expediates diagnosis. Working from downstream to upstream, analysts have a list of potential drivers to check. [14]

Similarly, this cause-and-effect schema also informs potential mitiga-tion measures and their limits. One of the tenets of production theory is that in the short run, there are production factors (e.g., a manufacturing plant) that determine a maximum output level because they cannot be changed in the short run. No matter how many other inputs (e.g., more workers) are thrown into the production process, one cannot exceed this fixed factor.[15] An example of a fixed factor in readiness is the number of units that can train at the large ranges that allow full units to train using live ammunition. The US Army, for example, only has two of these ranges —Combined Training Centers (CTCs). Brigade Combat Teams (BCTs) train one at a time in each of these CTCs, severely limiting the number of BCTs that the army can train in a year. The important information for readiness managers is that CTCs are a real limit on the ability to rapidly escalate the production of fully trained BCTs. Managers need to identify these fixed factors and manage production lines accordingly.

Sometimes diagnosis is difficult because the effects are masked. This can happen if, for example, a reduction in one driver or input is coun-terbalanced by an increase in another. If this is a conscious choice or is at least understood and monitored, there is no problem. However, if this is not the case and the unconscious mitigation abruptly stops, then it will be tricky to diagnose the original problem.

Figure 2. Direct Effects on the Readiness of a Given Unit or Organization

Readiness Element/Capability	Examples of Direct Causal Factors
A Combatant Commander ready to perform one of its assigned operational plans	• The readiness of assigned forces to perform that OPLAN • The probability the Services can generate allocated units that are ready to perform that OPLAN • The probability the Services and functional combatant commands can sustain deployed capabilities until plan objectives are met
A deployed or deployable long-distance strike unit ready to conduct its assigned mission	• Billets filled with qualified squadron personnel • Inventory of equipment (e.g. aircraft) capable of performing the assigned mission • Individual and unit training requirements in support of assigned mission
An aircraft depot that produces mission capable aircraft for units so they can meet deployment and pre-deployment requirements	• Depot billets filled with qualified maintenance personnel • Inventory of equipment (e.g. test benches) capable of meeting timely throughput requirements • Inventories of consumable and repairable spare parts
Inventory of qualified maintainers able to generate required throughput at depot-level maintenance facilities	• Recruiting and retention of qualified maintenance personnel • Throughput of individual-level training

Downstream

Upstream

Source: Laura Junor, "Managing Military Readiness," *Strategic Perspectives* no. 23 (National Defense University Press, February 2017), https://apps.dtic.mil/sti/pdfs/AD1030355.pdf.

A phenomenon called equipment cannibalization is a specific example of how problems can be hidden in equipment pipelines. Cannibalizations are a common unit-level reaction to problems in the equipment pipeline where some platforms (e.g., aircraft) are sacrificed to create a local spare-parts inventory to be used to keep other aircraft flying. While cannibalizations provide some buffer from a flawed supply pipeline, they can also create problems that range from accidental damage to parts due to excess handling to masking serious degradations in the supply chain. This was so much of an issue during the 1990s that Congress required the services report every cannibalization action in the Quarterly Readiness Report to Congress (QRRC).[16] Holding other things constant, the typical unit's drive to maintain readiness at all costs means that it could take several months or longer for these impacts to accumulate enough of an effect to significantly restrict the number of ready units and affect ongoing operations. While the direct operational and unit effects seem to be the most startling, its degradations in the upstream pipelines that cause the most damage in the long run. They are undoubtedly the most important tool for proactive readiness management.

The Problem with Manipulating Readiness Cycles and Tiering as Management Strategies.

The issue of readiness cycles and tiering as techniques for managing readiness have been around for a long time and are frequently identified as a means of protecting readiness while saving money. One idea is to lengthen readiness cycles so that units stay in a planned "unready" period until just before they are operationally needed. Another idea is to keep some units ready, and others perpetually at a managed less-than-ready status until they are needed. The common idea is derived from the realization that every deployed unit experiences some kind of readiness cycle that coincides with its deployment cycle. Units who are by all indications ready during their deployment come home and reset their equipment and disburse a significant portion of their personnel either for rest and recovery or to send them to schools. In most cases, a unit's readiness falls to some degree for some period just after a deployment and that may not necessarily be a problem. Periodically, budget-conscious leaders will propose that those unready, inter-deployment periods of readiness be prolonged for some units to yield savings that could be applied for other defense priorities. In fact, Richard Betts notes that a "country is militarily ready as long as the time needed to convert potential capability into the actual capability needed is no longer than the time between the decision to convert and the onset of war."[17] Essentially, he is saying that as long as units can recover their readiness completely before they next deploy, then there is no harm from the decline. Betts makes the case that allowing for such a decline is efficient, but only so long as the recovery criterion is met—and that is where things become challenging.

Managing the depth (how unready units would be allowed to go) and duration (how long they would remain unready) of these cycles to ensure that we do not deploy unready forces requires information about the limits of individual and unit skill recovery, dependable equipment management, and dependable deployment timelines—a very tall order.[18]

THE DEMAND FOR READY FORCES

The previous section presented a framework for understanding the systemic nature and causes of readiness deficiencies, but that is only half of the information necessary to adequately manage status. On the demand side, readiness management requires a clear articulation of prioritized threats and operations that the US military is expected to execute and the timelines they would likely employ. Readiness is contextual; without a framework for articulating what forces should be ready for and when they should be available, it is nearly impossible to determine whether the current supply of readiness is adequate or whether it jeopardizes national security. Betts argues just this point when explaining that "ready for what" and "ready when" are fundamental questions.[19] The nature of readiness management then involves comparing the schedule of the supply of ready forces to the schedule of the demand for those forces for the variety of operational demand signals that frame the National Defense Strategy. In turn, readiness managers must be able to identify supply-demand gaps along those schedules, the associated consequences, and the array of mitigation choices, including leaving the gap unfilled —all in fairly specific terms.

There are two general categories of demand for forces: 1) the rotational demand for presence and ongoing operations and 2) the demand for contingency operations. Rotational demand is far easier to articulate. It is the myriad of operations as well as presence, ally, and partner activities that employ the forces today.[20] The demand signal for contingency operations is not as well defined. Contingency demand represents the operations that the administration requires the department to be ready and notes in strategic documents such as the NDS. To assess the magnitude and consequence of deficiencies, the contingency demand must detail requirements for at least major capabilities along timelines. This demand signal could take the form of mature operational plans (OPLANs) or analytic scenarios.[21] Together, these two demand signals along with simultaneity assumptions form the essence of the operational requirements in national defense strategy.[22]

Consider various demand signals as alternate lenses with which to evaluate the adequacy of supply. Ideally, readiness managers would evaluate the current supply articulated in the assessment of readiness against the variety of rotational and contingency demands that comprise the NDS. With the correct tools, this approach allows users to consider variations in a single operational plan or varying degrees of simultaneity to explore force-generation limits. The ability to run iterations using real, systemic readiness data can be a powerful tool to evaluate investment and force-management options using defendable risk analysis.[23]

WHEN SUPPLY DOES NOT MEET DEMAND

The penultimate part of this readiness assessment framework is to assess supply or demand gaps. Readiness managers should expect gaps at this stage; there are highly complex pipelines, with exogenous forces like the civilian employment market, equipment failures, and unpredictable weather, and the demand signals (current and contingency) are typically far higher than the force can meet or sustain. Some gaps are easily mitigated; some are not. What is important is to be able to rationally choose what risk is acceptable and where scarce resources should be focused.

Much like the services' responsibility to forecast and articulate the consequences of deficiencies in their force-generation pipelines, the DoD must be equally clear about the expected consequences of not having individual requests for forces filled. These consequences should not only be logically consistent and defendable, but they should also explicitly consider how to manage the risk of any resulting future force generation constraints (e.g., presence gaps) that would result from failure to fill demand now. This is admittedly extraordinarily difficult to do, but it is an inherent requirement for a theater command before requesting additional resources. Unfortunately, there is not a lot of empirical work that estimates the value of overseas presence and engagement, either in terms of global security or global economics, leaving readiness and force managers with little more than anecdotes to consider. Recent work has begun to fill this information void by estimating the impact of reductions in US external security commitments on various US economic concerns and the prevalence or intensity of civil conflict abroad.[24]

ASSESSING RELATIVE RISK AND MITIGATION OPTIONS

Once a degradation is identified and its consequences determined, the next step for readiness managers is to determine the complete range of mitigation options. Each potential mitigation option should include a description of how it fixes the degradation, an estimate of how long that would take, and how much it would cost. One option should always be to do nothing other than monitor the situation, but that strategy must include a forecast of the consequences in the longer run should the degradation persist. Some mitigations involve supply-side considerations (e.g., invest resources differently) while others involve the demand side (e.g., accepting risk or reducing the demand by some means). In either case, readiness managers must specify the opportunity cost or consequences of the individual mitigations so that leadership can determine how the consequence of the mitigation compares with the consequence of the original degradation. On the supply side, it is conceivable that some mitigations may crowd out other critical investments or even the opportunity to resolve other readiness problems. On the demand side, there may be operational or strategic consequences to altering the demand signal. In either case, such information is important for leadership as they determine what to mitigate and how.

Reducing valid performance standards is never a viable mitigation strategy. An informed decision to accept risk is defendable, whereas quietly reducing standards without a deliberate risk assessment is not. The challenge here is that reducing standards tends not to be an overt reaction to a persistent readiness problem. Rather, it happens slowly and incrementally over time if only as a frustrated reaction to repeatedly missing seemingly impossible standards. It lends itself to the proverbial boiled frog syndrome, only in this case the outcome is forces that cannot execute a mission that they were purportedly ready to perform.[25] One common variant of this tendency is the understandable conflation of assigned and designed mission assessments for individual units.

For over a decade, BCTs were regularly performing counterinsurgency (COIN) missions in Iraq and Afghanistan. For many of the soldiers in these units, this was the only mission they knew. However, it was not the principal contingency mission the units were expected to engage in. But after decades of COIN-focused deployments, it was not a surprise to learn that the unit leadership had abandoned reporting on their ability to conduct their designed (non-COIN) mission in SORTS and only reported on their current assigned mission. This distorted the critical signal of the atrophy of full-spectrum capabilities required for other anticipated contingency missions. Once army leadership saw this shift in reporting, they swiftly and completely fixed it. Readiness managers and leadership alike must actively guard against this tendency to abandon assessing and reporting on their intended missions.

CONCLUSION

The analysis in this chapter is complicated and would benefit from innovative modeling and simulation toolsets. With advances in data analytics and associated breakthroughs in artificial intelligence, this approach will become more accurate and easier to execute. Until then analysts and mangers will be well served with more rudimentary efforts to capture data and analyses already available on the supply and demand for military capabilities. While this sort of analysis is difficult, it is possible. Mid-level managers in the DoD are expert at their parts in these production lines, and transaction-quality data are available in most cases. Furthermore, many of the relationships between inputs/ drivers and outputs have been studied by think tanks for years, so these processes are well understood. Any novelty here is about capturing this expertise for use in explaining changes in downstream readiness production. The more regular and detailed that link, the more control readiness managers will have.

NOTES

1. The exact date and origin of this proverb are unknown. It appeared in its earliest written form in the writings of Freidank, a thirteenth-century Germanic poet. A version was also recorded in English by John Gower in the late fourteenth century. For more information on the various incarnations of this proverb, see "Tales from the Nursery Rhyme, Part Twenty-Eight," https://windowthroughtime.wordpress.com/tag/richard-3rd-as-origin-of-for-want-of-a-nail/.

2. JP 1-02, *DOD Dictionary of Military and Associated Terms*, as of June 2020, https://www.jcs.mil/Portals/36/Documents/Doctrine/pubs/dictionary.pdf.

3. U.S. House of Representatives, Michele A. Flournoy, "DOD's Role in the Competition with China," testimony before the House Armed Services Committee, January 15, 2020, https://armedservices.house.gov/_cache/files/4/4/44fbef3d-138c-4a0a-b3a9-2f05c898578f/0E4943A5BFAEDA465D485A166FABCF5F.20200115-hasc-michele-flournoy-statement-vfinal.pdf.

4. Betts, *Military Readiness*, 24.

5. *Summary of the National Defense Strategy of the United States of America*, 2, https://DoD.defense.gov/Portals/1/Documents/pubs/2018-National-Defense-Strategy-Summary.pdf

6. Mark F. Cancian, "U.S. Military Forces in FY2020: The Strategic and Budget Context," Center for Strategic and International Studies, September 30, 2019, https://www.csis.org/analysis/us-military-forces-fy-2020-strategic-and-budget-context.

7. There are many variations of this old proverb, such as "For the want of a nail the shoe was lost," "For the want of a shoe the horse was lost," "For the want of a horse the rider was lost," "For the want of a rider the message was lost," "For the want of a message the battle was lost," "For the want of a battle the kingdom was lost," ending with "And all for the want of a horseshoe nail."

8. The most downstream readiness metric is a combatant commander's assessment of their ability to perform assigned and ongoing operations and the subordinates' assessments of their own abilities to perform the specific tasks that comprise those operations. These assessments are part of the Defense Readiness Reporting System (DRRS). For more detailed

of this and other traditional, downstream readiness metrics, see Junor, "Managing Military Readiness," Appendix A.

9. For a literature review on the effects of maintenance and personnel metrics on different aspects of readiness, see Junor, "Managing Military Readiness," Appendix B, 36–42.

10. Ibid., 37. Two studies find that the quality of personnel, independent of quantity, was a significant readiness driver. Quality in these studies was defined as the first principal component of five metrics: 1) percentage of the unit with a high school degree, 2) percentage of the unit testing in the upper mental group on the Armed Forces Qualification Test, 3) experience measured as the average length of service in the unit, 4) the percentage of the unit demoted within a given quarter, and 5) the frequency of rapid advancements. The techniques and definitions inherent in this are found in Junor, "Managing Military Readiness," Appendix B.

11. Ibid.

12. Ibid.

13. Service and joint units report downstream readiness assessments using a secure information system called the Defense Readiness Reporting System-Strategic (DRRS-S). This platform contains two complimentary ways of reporting readiness. One, the Status of Resources and Training System (SORTS), is input or resource based. It is primarily focused on service units (ships, battalions, and squadrons) and answers whether these units have the requisite personnel, supply, equipment, and training to conduct their missions. The other, Defense Readiness Reporting System (DRRS), is output or mission/task based. It includes joint units up to combatant commands and asks commanders to assess what tasks and missions the unit can perform. Together these assessments provide a downstream view of a unit's capability status. For a more complete discussion, see Laura Junor, "Managing Military Readiness," Appendix A.

14. There are several options for quantifying effects of changes in readiness inputs/drivers on output measures. Some processes, like sortie generation, have been modeled using various statistical and simulation tools for years. Reasonable cause-and-effect estimates are quite possible. In other less-studied areas, the production managers provide informed estimates. Over time, with advances in data analytics, these estimates will be easier to make and be more accurate.

15. In fact, if you throw in too many other factors (e.g., workers), you create what is called diminishing marginal returns in production theory. In our example, this would happen as the manufacturing plant becomes

too crowded with workers and every additional worker reduces output. Understanding the role of fixed factors is an essential part of monitoring healthy production pipelines.

16. The Quarterly Readiness Report to Congress (QRRC) is a congressionally mandated report that provides both volumes of readiness measures (e.g., cannibalization rates) and downstream metrics such as assessments by each combatant commander of the "readiness of the command to conduct operations in a multidomain battle that integrates ground, sea, air, space, cyber, and special operations forces". For more details, see https://www.law.cornell.edu/uscode/text/10/482.

17. Betts, *Military Readiness*, 28.

18. For a more robust discussion of cycles and tiering see Junor, "Managing Military Readiness," 15–19.

19. Note that the supply framework from the last section can describe what readiness capabilities are being produced and predict when they will be ready.

20. For a more comprehensive discussion of rotation demand, see Junor, "Managing Military Readiness," 21–24.

21. For a discussion and history of force planning scenarios, see Larson, *Force Planning Scenarios, 1945–2016.*

22. Leaders recognize that planning for many major operations occurring at the same time would require far more resources than the nation would responsibly provide given the relatively low probability involved. Typically, defense strategies require that the force be capable of conducting two major operations with a specified degree of overlap and outcome expectations. Simultaneity assumptions reflect those expectations. Obviously, decisively winning two operations in different theaters at the same time represents a tougher challenge than winning one while achieving a more modest objective for the second.

23. Together the range of these assessments indicate the limits of current readiness and the risks associated with those limits. This risk analysis will inform decisions about when and how to mitigate readiness shortfalls.

24. One example that helps clarify the value of overseas presence is Flournoy and Davidson, "Obama's New Global Posture."

25. The refers to the anecdote about a frog being slowly boiled alive. The premise is that if a frog is placed in boiling water, it will jump out; but if it is placed in cold water that heats up slowly, it will not perceive the danger and will eventually be cooked to death. The story is often used

as a metaphor for the inability or unwillingness of people to react to significant changes that occur gradually.

CHAPTER 10

PROGRAMMING DEPARTMENT OF DEFENSE STRATEGIC PRIORITIES

John Ferrari

In an ideal world, the defense budget debate would be fact-based and informed. Of course, no one should be under any illusion that final defense budgets spring fully formed from dispassionate strategic analyses. The last word on defense spending always results from the clash of strategic thought and the reigning political environment.

—Lieutenant General Thomas Spoehr (retired)[1]

INTRODUCTION

Previous chapters in this volume have provided a context through which to understand the Department of Defense process for allocating resources to fulfill military strategy. They have also illustrated how the US constitutional structure and budget process render holistic planning difficult. Nonetheless, leaders must make strategic choices for the US military in terms of the four pillars, which were laid out in chapter 8: force size and composition, force posture, force modernization, and force readiness. And to be successful, they must also be able to accurately identify and evaluate trade space among the choices as well as the risks entailed.

Despite all the efforts at prognostication, attempts to predict the nature and location of the next war have been dismal failures. In 2011, former Secretary of Defense Robert Gates observed that "we get it wrong 100% of the time."[2] The following year, General James Mattis, then serving as the US Central Command (CENTCOM) Commander, testified before the Senate Armed Services Committee, stating "I think, as we look toward the future, I have been a horrible prophet. I have never fought anywhere I expected to in all my years."[3] Simply put, "the enemy gets a vote" and can therefore not be expected to conform its tactics and operations to US planning assumptions. In spite of this abysmal forecasting track record, planning and resourcing must take place, and they have to be optimized to the degree possible to provide capabilities to fulfill the requirements of the extant strategy. The strategy, however flawed, drives resourcing choices and the resulting trades and risk.

The strategic resourcing process within the DoD is called "programming," and it is the second part of a macro-level process called Planning, Programming, Budget, and Execution (PPBE) that was put in place by then Secretary of Defense Robert McNamara in the 1960s.[4] A sequential process designed by the "whiz kids," the output of one phase of the PPBE process is intended to serve as the input to the next phase. The output of the planning phase, which constitutes the overall DoD strategy, is then intended to be the input to the programming phase.[5] The output of the programming phase, the "program," which is referred to as a Program Objective Memorandum (POM), serves as the DoD input for creating the President's Budget (PB) that is sent to Congress each February.[6] The President's Budget serves as the input to Congress, which then produces the appropriations that drive the execution phase of the PPBE process. In theory, the strategy drives the POM, which drives the budget, which drives execution. If only it were that simple. This purpose of this chapter is to provide insights for aspiring strategists and policy makers within the DoD on how to better ensure that strategy actually drives resourcing —because in practice, this is often not the case.

The four phases of PPBE are managed by different groups of people, each with different skill sets, bureaucratic interests, and timelines. The planning phase is managed by strategists and policy makers who endeavor to affect the global geopolitical environment to further broad strategic objectives. The programming phase is led by programmers, who operate within a zero-sum scramble for dollars and whose job it is to make the "math" work. The budget phase is led by financial managers whose focus is to convince the appropriators in Congress to accept the President's Budget proposal; think of them as the President's "lobbying" force. The execution phase is led by accountants—the "bean counters"—whose focus is to make sure the dollars are spent exactly as they were appropriated. This chapter is principally focused on the relationship between the strategists and the programmers.

If the strategists and programmers are aligned, the strategy, as envisioned by its authors, will flow smoothly through DoD and OMB to Congress. If the strategists and programmers are not aligned, the programmers will have the greater say in how the resources are allocated because of their position further downstream in the PPBE process. In order to implement a strategy, programmers must allocate resources to support it. However, programmers do not need a strategy to allocate resources. Therefore, strategists must fully understand, embrace, and participate in the programming process to ensure their strategy is actually resourced. Without sufficient advocacy, a strategy is merely a piece of paper, while the program is the official database of what will happen, and as the saying goes, one should always "follow the money," which will inevitably take precedence.

BOUND BY TIME AND RESOURCES
[SUCCESSFUL STRATEGY =$T^4 * R^2$]

Within the programming phase, there are two critical variables: time and resources. The programming phase is a yearly cycle of time, which allocates resources in five-year time increments and is bound to a fifteen-year national time cycle of defense resourcing. The overall success of the process is dependent upon how much *time* strategists dedicate to learning and participating in the PPBE process itself. There are hundreds of thousands of individual resource decisions that cumulatively become the de facto defense strategy. If future policy makers take away only one thought from this chapter, it should be the following: *If policy makers want to drive resourcing, then the strategy must be bound by both time and resources.*

These four dimensions of time (planning, programming, budget, execution) and two dimensions of resources (strategy and allocation) are the primary obstacles to effective strategy implementation for DoD. Overcoming them falls squarely on the shoulders of those who write the strategy because a programmer can produce a program without a strategy, but a strategy cannot be implemented without the programmer due to the sequential nature of the PPBE process.

The importance of the information asymmetry between strategists and programmers within the DoD cannot be overstated. To work in the system, programmers must understand it. This is not the case for policy makers and strategists, which puts them at a distinct disadvantage relative to the programmers. To become effective inside the PPBE process is an enormous time sink, because the strategy/resource decision space is both multidimensional and complex. Additionally, the knowledge return versus time inputted is not linear, it is in fact exponential—the more time an individual invests, the more they understand how to influence the process; mastering this domain requires an enormous time commitment and therefore rewards multiyear participants. There are no shortcuts. The barrier to entry into the programming process is nonnegotiable— the entrance criteria is based solely upon how much personal time an individual is willing to invest in the learning the process.

PROGRAMMING IS UNIQUE TO DoD

Programming constitutes the bridge between the strategy (what DoD wants to do) and implementation (what DoD is funded to do). Most executive branch departments within the US government go straight from strategy to budget, but former Secretary of Defense Robert McNamara instituted the programming phase during the 1960s because the strategy and budget were not aligned.[7]

In his book *The Power Broker*, Robert Caro explains how Robert Moses built the infrastructure of modern-day New York over the course of four decades.[8] Caro describes how Moses would get legislators to fund the first few miles of a highway, knowing that once ground was broken, the rest of the money would follow because nobody wanted to see a hole in the ground.[9] During the 1960s, DoD budgeting was in a similar situation. The military services would start huge modernization programs in a single-year budget, without worrying how much they cost to complete.[10] Thus, the DoD could write an expansive strategy, put initial funding in place, and then assume more money would follow. Programmers were tasked to stop this from happening. Therefore, the entire PPBE process is built around the programming phase with the programming process itself functioning as Secretary McNamara's enduring legacy.[11]

The programming process is designed to prevent the military services from buying the "first mile of the highway" in a single budget year by balancing resources for the following five-year time increment. Within the DoD, military departments assemble the hundreds of thousands of data elements, called the POM, or Future Years Defense Plan (FYDP), which encompasses a five-year time horizon.[12] The first year of the POM, is "peeled" off at the end of the process and sent to Congress as the budget. The five-year time frame is intended to prevent the "unfinished road" within the DoD as a whole.

The Yearly Cycle of the Process (T1)

To work effectively within the PPBE process, one must understand multiple dimensions of time. The first dimension is the tyranny of the POM process. The POM process, as designed by systems engineers, is a disciplined, data-driven, time-based process, tied to delivering its output in time to enable the annual budget submission in February. Within the PPBE process, certain events happen at certain times; if one does not show up to the right meeting, with the right information, at the right time of the year, it is like missing a train that only comes annually. Once the train leaves the station, it does not return for another year. Therefore, if the strategy submission is not synchronized to the PPBE phases, the strategy will be irrelevant.

To maximize the value of a strategy, leaders and policy makers within the DoD should deliver the strategy to programmers by October or November. If the strategy document is delivered past November, the proverbial train would have already left the station, and by the following year, the document will likely have become irrelevant. To make this already complicated process more complex, it is actually a series of four overlapping one-year cycles that takes forty-eight months to elapse. Think of the PPBE process not as a single-year train schedule but rather a four-year train schedule, with four trains running all at the same time. As a strategist, one not only has to know what time to show up to the station but also to which train to hook the strategy document—or the strategy will likely wind up on the wrong train.

Figure 3 shows how the yearly PPBE process is actually four overlapping phases that together take forty-eight months to complete. A policy maker who can master this timeline will be able to write for, communicate with, and influence resourcing within the DoD. So, how does a one-year process take forty-eight months to finish? It is driven by the constitutional responsibility of Congress to raise an army and maintain a navy.[13] Of the forty-eight-month cycle, Congress controls the last twenty months.

Figure 3. The 48-month PPBE Process.

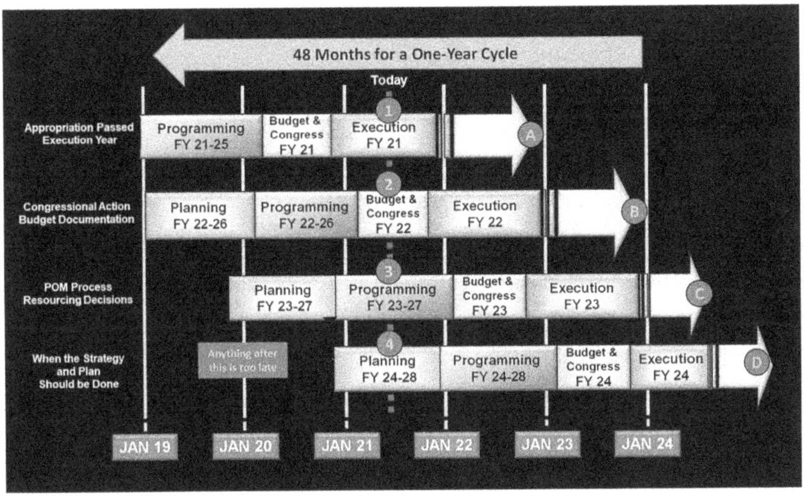

Source: Adapted from *How the Army Runs* by the Army Force Management School, 3–36, https://publications.armywarcollege.edu/pubs/3736.pdf.

To better understand the four phases, they will be explained in reverse order. Figure 3 depicts four train tracks, A through D, with each train in a different phase of the PPBE process. For example, Train A is in the execution phase of 2021, while Train D is in the planning phase for FY24–28. For explanatory purposes, let us assume that it is the summer of 2021.

Track A is the execution phase of the PPBE and is depicted at point 1 on the diagram. This takes twelve months of the cycle (from October 2020 through September 2021 on track A). This is the phase in which the DoD actually spends money, and it is led by the accountants who are enforcing the enacted budget appropriations; there is little to no flexibility in this part of the PPBE. The cycle is locked in progress. Changes require congressional approval.

Track B is the budgeting phase of the PPBE and is depicted at point 2 in figure 3. This encompasses nine months of the cycle from January 2021 through September 2021. The financial managers are the dominant participants in this part of the cycle; and in this phase, they are preparing, submitting, and defending the President's Budget to Congress. There is little to nothing that a change in strategy can influence in this part of the cycle.

Track C represents the programming phase of the PPBE and is depicted at point 3 in figure 3. The programmers dominate these fourteen months of the cycle from November 2020 through December 2021. It is the period in which the programmers make their resource decisions. The programming phase of the cycle is longest, probably because it was the programmers who put the PPBE into place. However, ownership of this phase is split between the individual services and the Office of the Secretary of Defense (OSD). The military services are the dominant participants from November through June (eight months) each year, then the OSD dominates the time from July through December (six months). Once the programmers start their process, it is difficult to change strategic direction—think of it as trying to show up with a new house plan after the foundation has been poured. If you already had a strategy written, this is the train you would board.

Track D is the most important part of the process for a strategist or policy maker, it is the planning phase of the PPBE depicted as point 4 in figure 3. This is the ten months of the cycle from January 2021 through October 2021 in which the strategist or policy maker can have the most influence and flexibility. This is the period in which a new strategy should be written.

One must understand that at any given time, all four trains are running. It resembles a four-dimensional chess game of resourcing and flexibility. The services program runs from November through June of each year on Track C. During this time, they build their programs in accordance with their priorities; that is, if one exists and is executable, with a nod towards the DoD or service strategy. In June of each year the services submit their programs to the OSD, and from July to December they defend their resource decisions. The word "defense" is critical. This part of the process is not collaborative. If a strategy supports existing resource decisions, it will be embraced, but if the strategy calls for a change in resources, the programmers will reject the strategy and continue to make resourcing decisions.

In certain respects, it takes four years to fully implement a strategic change. The strategy document needs to explicitly take into account and give guidance on what can realistically be done in each of the phases and fiscal years. Being specific in the strategy document will increase the chance the strategy will be implemented as soon as possible. For example, if a strategy document outlines changes that can be made without congressional approval, in order to set the stage for larger changes later, these changes must be both few and explicit.

The Five-Year Horizon of Programming Decisions (T2, R1)

This section explains the second dimension of time (the five-year programming horizon) and the first dimension of constrained resourcing (the iron triangle). It is intended to enable policy makers to better understand how their documents can more effectively influence programmers, and it also recommends that strategy documents conform to the programming culture of resource decision-making. It is important to remember that the entire PPBE system is designed to prevent the building of the first mile of the highway without knowing where the rest of the money is coming from. Therefore, the programmers' paramount concern is balancing military manpower, modernization, and readiness over the next five-year period. Taken together, these three variables constitute the iron triangle of the DoD, which was discussed in detail in the preceding two chapters of this book.

The programming phase is dominated by the five-year program called the Future Years Defense Program (FYDP). Figure 4 depicts the seven years that are always "at play," of which only five years are part of the programming process. The first two years, those of execution and budget, are as previously discussed, assumed as fixed and unchangeable by the programmers since they are in the hands of the US Congress. Even the Secretary of Defense needs congressional approval to make all but minor changes to resourcing. This leaves the next five years, as shown in figure 4, FY23–27 to the programmers.

Figure 4. The Five-Year POM Timeline.

FY 21	FY 22	FY 23	FY 24	FY 25	FY 26	FY 27
Execution	Budget at Congress	POM Years Open for Decisions				

Source: Adapted from *How the Army* Runs by the Army Force Management School, https://publications.armywarcollege.edu/pubs/3736.pdf.

Within the DoD, some may have heard the saying "all it takes to make a new strategy is a few hours and Microsoft Word." This is flippant, but it is built on the notion that it takes decades of patience and consistency to build an effective military force and that constant change fritters away resources that ultimately reduces overall military capability and effectiveness. Policy makers must realize that programmers have to balance the resources within the iron triangle in order for the DoD to function. The essence of this is the balance among the variables of manpower (which takes years to grow and train), modernization (which takes decades within the acquisition process), and military readiness (which is hard to define, even harder to build, but easy to destroy). The majority of the $700B+ yearly defense budget ($3.5T over the five years) is fixed, and change is costly. Strategies that inject change, without significant guidance on cost savings, are likely to be ineffective. If policy makers want to effectively drive resourcing, then the strategy must be bound by both time and resources. In this case, the strategy must be bound by the five years of available resources.

The five-year POM is the summation of every strategy and decision (good and bad) that has come and gone in the past sixty years. Therefore, change is slow and incremental at best, and any strategy that proposes sweeping changes and does not account for how much money is needed will almost certainly fail to have any impact. Another key point is that strategies written for time periods beyond the five years in question are also less relevant. There are three reasons why it is hard to move resources among the three points of the iron triangle. First, it is extremely costly to do so; therefore, strategic realignments among the points of the triangle have significant cost implications, and the cost of the change may exceed the benefit. Second, each point of the triangle has a constituency, so rebalancing brings external stakeholders into the debate, thereby complicating the process further with the need to seek consensus. Finally, each point of the triangle exists on a different temporal timeline—readiness exists in the near term, manpower in the midterm, and modernization in the long term. Trading capability, costs, and benefits across time is fraught with risk and rarely achieves the desired outcomes.

It is thus clear that it is as imperative for policy makers and program-mers to work collaboratively as it is difficult for them to actually do so. To improve the necessary collaboration, improved understanding of the process is necessary on both sides of the "strategy/programming divide." A first step toward this is that strategists should write the strategy documents to conform to the programming culture of decision-making. This needs to be done both in terms of the five-year period of the POM as well as the aim of maintaining balance within the iron triangle. Second, policy makers should purposefully write up strategy documents with the military service programmers as the intended customer. Getting military service buy in upfront as well as understanding the friction points and acknowledging that the services control the next step of the PPBE process after the strategy is complete are critical. Lastly, it is crucial to understand that the strategy and resourcing within DoD must be integrated into the broader national security apparatus.

This requires three partnerships between programmers, policy makers, and strategists who sit at different levels within the PPBE process. In all three cases, it is the strategy writers who must establish these partnerships because the programmers can accomplish their mission alone. However, when this happens, the strategic direction of DoD suffers. Therefore, those responsible for the planning phase of PPBE must incorporate the programmers into their writing process and become active participants in the programming decisions.

The first of the three necessary partnerships exists at the OSD level —the strategists of OSD policy and the programmers of OSD Cost Assessment and Program Evaluation (CAPE)[14] must work in tandem to craft and program the DoD's strategy. OSD policy and CAPE strategists do already partner reasonably well together, but this collaboration must be strengthened.

The second of these partnerships would team the NSC staff and OMB staff to interact with the entire national security establishment with regard to resourcing. Right now, there are many policy and strategy documents at the interagency level, but few, if any, resourcing documents. What is needed is a national security strategic programming guidance document that is the product of a national-level planning phase that provides specific guidance to drive resourcing.

The third of the three partnerships would integrate the service planners and programmers into a combined organization. While typically the planners are in the strategy and operations part of the service staff (in military parlance, they would be part of the "3" or "5" office), the programmers are normally in the "8," which is responsible for programming.[15] Given that the services are the only organization that actually produce a POM, this teaming up of service strategists and programmers would have an immediate and beneficial impact in ensuring timely strategy that informs the balancing of the iron triangle.

THE NATIONAL CYCLES OF FUNDING THAT OVERWHELM STRATEGY (T3, R2)

The third dimension of time and the second dimension of resourcing is the cyclical nature of national funding. The national defense funding for the United States is a sine wave that goes up during war and goes down during times of budget stringency related to the guns versus butter national strategic debate as depicted in figure 5. Since World War II, there have been five distinct peaks in funding national defense programs. The peak-to-peak cycle occurs over roughly a fifteen-year period, with the buildup phase (trough to peak) lasting an average of 8.8 years and the average budget stringency lasting an average of 7.5 years.[16]

It is ironic that the one thing that the defense "sine wave" budget does not conform to is the written DoD strategy. If the DoD or service strategy does not fit the national budget cycle, the strategy will at best be ignored and at worse lead to hollow forces.[17] This dilemma boils down to the age-old question: "Does strategy drive resourcing or does resourcing drive strategy?" As figure 5 depicts, only the strategic decision to go to war (whether hot or cold) drives resources; a great strategic plan has never driven resourcing. As Dr. Frank Hoffman, the principal author of the 2018 National Defense Strategy, explains that strategists should understand their limits, noting, "One makes strategy and executes it in the real world, and the real world is an environment in which constraints are almost always operative. The most obvious of which are time, information, and resources."[18]

Figure 5. The National Resourcing Cycle.

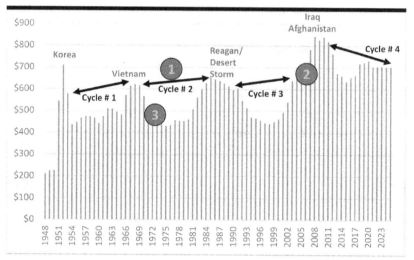

Source: Adapted from Table 6-1, "Department of Defense Total Obligation Authority (TOA), 1948–2025 in FY 2021 Constant Dollars," Office of the Undersecretary of Defense (Comptroller), *National Defense Budget Estimates for FY 2021*, p. 80.

Next, we will examine the larger forces at play in defense strategy, which will further underscore how important it is for policy makers and planners to write a strategy that conforms to the resource cycle. In finance, there is a saying "do not fight the Fed"—the same holds true here.[19] Do not fight the resource cycle; instead write a strategy that will help the programmers to allocate resources. Wars are the singular predictive event that drive resources, but nobody ever goes to war because it was in a strategic plan; more often than not, war happens. Absent war, resources are either dropping or being held constant at a level below what the DoD believes is adequate. What is never known is the length of time between wars and the full depth of the downturn in resources. When war arrives, resources flow, leading strategies to expand. Wars eventually end, and resources contract. A well-written strategy document understands this cycle and serves as a guidepost to operate within it.

So, how then does one write a strategy that prepares the DoD and the military to be better positioned to expand and contract with the cycle?

First, based on the author's experience, it is impossible to predict or influence the resource cycle. A strategy that does not take this into account will be unable to adapt to unforeseen consequences, and this failure will be paid for with lives. As many books have documented, over the past one hundred years, America has been caught unready rather than prepared for war.[20] From the Battle of Kasserine Pass in World War II to the "hillbilly armor" in Iraq, American strategies have resulted in unreadiness and hollow forces. "You go to war with the army you have, not the army you wish you had"[21] is not the epithet associated with great strategic planning.

To avoid this seemingly inescapable trap, strategies must employ option theory, which is based on the assumption that firms have some choice in when to invest. For the DoD, investment in a military capability is similar to an option: there is an opportunity, but not an obligation, to use it. Two aspects of an option increase its value. The first is uncertainty; the more uncertain an outcome, the more valuable an option becomes. The second is time duration; the longer the time horizon, the more valuable the option becomes.[22] Therefore, a strategy, up front, should build in a series of options. The options will differ within the strategy, depending upon where in the resourcing cycle the strategy is being written.

If we find ourselves at the peak of resources (as is depicted at point 1 of figure 5), we should never assume that resourcing will remain constant. In fact, a good strategy would be preparing for the inevitable decline that is to be forthcoming. Assuming resources will remain flush is the worst flaw of the entire defense programming process. With nearly 100% certainty, this is a false assumption, which serves to undermine the entire program and strategy.

If we find ourselves with increasing resources, as happens every decade or so (as depicted at point 2 of figure 5), it would be prudent not to believe this situation will endure. It should be the job of the strategist "to take the punch bowl away, just as the party gets going."[23] With resources expanding, the impulse is to expand the force, increase readiness, and begin decades-long modernization programs—all assuming resources will remain constant or increase. All this accomplishes is to waste money and time because the inevitable downturn in resourcing will result in having to shrink the force, decrease readiness, and cancel modernization programs. When resources are increasing, a good strategy is one that serves as a brake on the system and prepares the force to remain ready when the increased resources disappear. The strategy at this point should be one planned with the awareness that "the party will eventually end."

If we find ourselves with decreasing resources, as is the case every decade or so (depicted at point 3 of figure 5), we should not precipitously decrease the size of the force or decrease readiness by assuming that the country will not go to war for a long time. Assuming a decade of peace in order to increase modernization and long-term investments almost guarantees being caught unprepared for the next war. A good strategy, during a period of fiscal stringency will protect the near-term readiness of the force, while building in modernization options and ramps for when resources are again increased. The strategy at this point should be planned with the caution that "war is coming sooner than you think."

The prevailing dogma within the DoD that strategy drives resources and that strategies need to define future force capabilities is thus the wrong way to approach the problem. A good strategy is one that embraces uncertainty and plans for it. A good strategy forces explicit decisions concerning the iron-triangle balance and has options, guided not by wants or desires but by the cold hard facts of dollars available and decisions to go to war. The bottom line is if the strategy wants to drive resourcing, then the strategy must be bound by resources.

THE TIME ON AN INDIVIDUAL'S DAILY CALENDAR (T4)

The fourth dimension of time relates to the amount of time it takes to become effective within the PPBE process. This section explains the information asymmetry that exists within the different phases of the PPBE and recommends that aspiring DoD policy makers study and embed themselves within the entire PPBE process. Programming is a full-time, hands-on commitment. To have true positive influence on the process requires investing hundreds of hours in preparation and working multiple jobs at the Pentagon. High rank and position cannot shortcut the process. Part-time programming may alone account for the dismal outcomes associated with America's first battles.[24] An individual's time is the barrier to entry to the programming process, and it is a very high barrier indeed. Whether one believes in the 10,000-hour rule or the seven-year rule,[25] one ultimately has to invest time to be effective.

The good news is that knowledge and expertise is cumulative and can be built over a number of years; the bad news is there is no substitute. The process of driving change in resourcing must be done top-down, but if those who write the strategy and the leaders who want to drive change do not invest the time, then the process defaults to a bottom-up decision-making pyramid where thousands of individual decisions are made regardless of what the strategy directs. Undoing those decisions, takes hundreds, if not thousands, of hours. Programmers at the bottom of the process pyramid have invested years into this process and are very good at protecting their funding.

The Future Years Defense Program (FYDP) consists of a database that contains hundreds of thousands, and cumulatively millions, of pieces of data from across the entire DoD.[26] It is both wide and amazingly deep. Each of the data elements has a sponsor whose sole job is to protect those resources. These sponsors represent the missions and functions that they believe are critical and can be justified by the current strategy. If the strategy shifts, their justification for resources will shift with it.

To overcome this challenge, strategists and policy makers need to spend time as programmers inside acquisition programs and understand how manpower is trained and educated. They need to do multiple jobs in the Pentagon in order to gain multiple perspectives. They also need to master a basic knowledge of economics, risk shifting, and option pricing, which will give them the tools of a strategist to put together a coherent narrative that others can follow. More importantly, leaders must enable the growth and advancement of a cadre of experts who are both excellent strategists and outstanding programmers.

The services each run technical schools that teach the Planning, Programming, Budgeting and Execution process as part of the curriculum. But there is no reference or course that fully captures the nuance and complexities of the strategy-resourcing relationship. This is especially challenging for government civilian employees and political appointees who likely have limited experience in the process and who, unlike their uniformed colleagues, do not have the opportunity for formal coursework at military schools. Nevertheless, there are several references that can provide a basic understanding of PPBE for the interested and uninitiated. Chapter 5, "Planning, Programming, Budgeting, Execution, the Department of Defense," in *Buying National Security* provides a good, if dated, overview.[27] Additionally, the US Army publishes *How the Army Runs*, which provides an extremely detailed explanation of every step of the PPBE process.[28] Finally, the Congressional Research Service publishes helpful summaries of the Future Years Defense Program and PPBE process, as well as of the DoD procurement process.[29] However, none of these references capture the complexities of actually making the system work. The best way to learn the strategy-resourcing relationship remains having the experience of working within it—and that takes time.

CONCLUSION: BOUND BY TIME AND RESOURCES
[SUCCESSFUL STRATEGY $=T^4 * R^2$]

There are no shortcuts; only by understanding time and bounding the strategy to available resources can a strategist be effective. A strategy can drive resource allocations, but only if it works effectively within the constraints of the decade-long national cycle of funding and is written for the five-year time horizon of resourcing decisions. A strategy not only has to be delivered at the correct time in the cycle, but it also needs to be written by someone who has invested the time to truly understand how the iron triangle of manpower, readiness, and modernization all have to come together to have an effective fighting force. The stakes are high—if the strategy fails, it is the 18-year-old soldier on the frontline that will pay the ultimate price.

NOTES

1. Thomas Spoehr, "Differing Views on the Defense Budget: Six Blind Men and the Elephant," *War on the Rocks*, Jan 14, 2021, https://warontherocks. com/2021/01/the-six-blind-men-and-the-elephant-differing-views-on-the-u-s-defense-budget/.
2. Secretary of Defense, Robert Gates, quoted in "100% Right, 0% of the Time," by Micah Zenko, *Foreign Policy*, Oct 16, 2012, https://foreignpolicy. com/2012/10/16/100-right-0-of-the-time/.
3. CENTCOM Commander, General James Mattis, quoted in "100% Right, 0% of the Time," by Micah Zenko, *Foreign Policy*, Oct 16, 2012, https:// foreignpolicy.com/2012/10/16/100-right-0-of-the-time/.
4. "Planning, Programming, Budgeting and Execution Process: DAU Glossary Definition," Defense Acquisition University, https://www.dau. edu/acquipedia/pages/articledetails.aspx#!154.
5. The "whiz kids" was a name given to a group of experts from RAND Corporation that Robert McNamara brought to DoD in the 1960s whose skills included economic analysis, operations research, game theory, and computing.
6. The POM is an acronym that is meaningless today. It stands for Program Objective Memorandum and dates back to days when memoranda were used within the DoD.
7. Matthew Fay, "Pentagon Planning and the Problem of Predictability," *The Niskanen Center,* May 27, 2015, https://www.niskanencenter.org/ pentagon-planning-and-the-problem-of-predictability/.
8. Caro, *The Power Broker, Robert Moses and the Fall of New York*, 526–540.
9. Ibid.
10. Fox et al., *Defense Acquisition Reform 1960–2009*, 35.
11. Matthew Fay, "Rationalizing McNamara's Legacy," *War on the Rocks*, Aug 5, 2016, https://warontherocks.com/2016/08/rationalizing-mcnamaras-legacy/.
12. FYDP stands for Future Years Defense Program. For a thorough explanation of the POM and the FYDP, see U.S. Library of Congress, Congressional Research Service, *Defense Primer: Future Years Defense Program* by Brendan McGarry and Heidi Peters.

13. Under Article I, Section 8, Congress has the power to declare war, raise and support armies, provide and maintain a navy, and organize, arm, discipline, and call forth a militia.

14. The direct successor to the organization that McNamara created, CAPE is the OSD organization that runs the programming phase and comprises the modern-day "whiz kids."

15. American military staff sections are identified by their number. For example, a "1" would generally be associated with personnel and a "3" would be associated with operations. For a breakdown of how this numbering system works, see *The Joint Office Handbook*, 4th ed., 2018, 34, https://wss.apan.org/public/jom_jqs/Shared%20Documents/Joint%20Staff%20Officer%20Handbook.pdf.

16. There have been five distinct "peaks" in defense spending associated with the Korean conflict, the Vietnam War, the Reagan administration buildup, Iraq/Afghanistan, and the Trump administration buildup. Average buildup was assessed as the number of years between trough to peak in defense spending, and average stringency was assessed as the number of years from peak to subsequent trough. The analysis assumes that the DoD budget peaks with the FY 2022 budget submission. Data is taken from Total Obligation Authority found in chapter 7 of "Tables of National Defense Budget Estimates," Office of the Under Secretary of Defense (Comptroller), https://comptroller.defense.gov/Budget-Materials/.

17. The term "hollow force" refers to a military formation that upon initial inspection appears ready but upon closer examination is uncovered to be unfit for combat operations due to deficiencies in training, equipment, funding, or personnel. The term was initially used to describe the U.S. Army in the 1970s. For further information, read Jack Miller, "Hollow Force: The Trade Off Between Readiness and Modernization, *Modern War Institute,* Aug 24, 2014, https://mwi.usma.edu/hollow-force/.

18. Hoffman, "Grand Strategy," 478.

19. Zweig, *Winning on Wall Street*, 97.

20. Heller and Stoft, eds., *America's First Battles*.

21. Eric Schmitt, "Iraq Bound Troops Confront Rumsfeld Over Lack of Armor," *The New York Times*, December 8, 2004, https://www.nytimes.com/2004/12/08/international/middleeast/iraqbound-troops-confront-rumsfeld-over-lack-of.html.

22. For more detailed information on "Option Theory," consult "Real Options Theory," *Economics Terms A to Z*, https://www.economist.com/

economics-a-to-z/r#node-21529847; and Leslie and Michaels, "The Real Power of Real Options."

23. Speech (October 19, 1955) by William McChesney Martin, who served as chairman of the Federal Reserve from 1951 through 1970, to the New York Group of the Investment Bankers Association of America.

24. From Kaserne Pass to Task Force Smith to Desert One to Operation Iraqi Freedom, the American military has consistently been caught unprepared for the first battle of the next war in spite of the efforts of strategists, programmers, and the entire national security establishment whose jobs are to ensure that this does not happen.

25. Popularized by Malcom Gladwell in his bestseller, *Outliers.*

26. For a detailed explanation of the FYDP, see U.S. Library of Congress, Congressional Research Service, *Defense Primer: Future Years Defense Program* by Brendan McGarry and Heidi Peters.

27. Gordon and Williams, *Buying National Security,* "Planning, Programming, Budgeting, Execution, the Department of Defense," Chapter 5, 93-119.

28. *How the Army Runs,* https://publications.armywarcollege.edu/publication-detail.cfm?publicationID=3736,

29. U.S. Library of Congress, Congressional Research Service, *Defense Primer: Future Years Defense Program* by Brendan McGarry and Heidi Peters, *Defense Primer: Planning, Programming, Budgeting and Execution* by Brendan McGarry, and *Defense Primer: Defense Procurement* by Heidi Peters and Brendan McGarry.

CHAPTER 11

RESOURCING
HOMELAND SECURITY

Mark Troutman

The only turf we should be worried about protecting is the turf
we stand on.

—DHS Secretary Tom Ridge[1]

MISSION, THREATS, AND HISTORY[2]

The Department of Homeland Security (DHS) is the newest cabinet-level
department in the American government. In 2002, Congress passed the
Homeland Security Act to create the department and fix the systemic
governmental failures that enabled the 9/11 hijackers.[3] The legislation
combined twenty-two previously separate departments and more than
a dozen smaller agencies into a single entity with the common mission

to safeguard the United States by protecting the systems that enable democratic and economic activity.[4] While the DoD secures the country from military attack, the DHS has the mission to defend against nonmilitary threats.[5]

To fulfill this broad mandate, the DHS must prevent terrorism, enhance and ensure infrastructure resilience, secure US borders, enforce immigration laws, secure cyberspace, and provide disaster response.[6] Although smaller than some other departments of the national security enterprise, the DHS is still huge. It employs over 240,000 people, and its 2021 federal budget request was $72 billion.[7] The DHS's mandate is complicated by the fact that many of its responsibilities involve the protection of assets and functions that are under the authority of state and local governments, the private sector, and even individuals.[8] As result, the department often finds itself in the position of having ultimate responsibility while lacking full authority.

Securing free and open societies is a challenge as systems built on individual liberty, free markets, and a federal governance structure present many vulnerabilities for malign actors to exploit. Thomas Jefferson famously articulated every American citizen's right to "life, liberty, and the pursuit of happiness," in the Declaration of Independence. Nearly 250 years later, the US government often finds these "inalienable" rights to be in competition with one another, if not, conflict. As a result, the DHS must ensure security while simultaneously safeguarding the freedoms and openness for which the United States is so famous and constitute a core national strength.

The DHS must also ensure resilience in the case that disaster—whether natural or manmade—should strike. Security and resilience are multifaceted terms that are both defined in President Barack Obama's Presidential Policy Directive 21 (PPD-21), which remains in effect. According to PPD-21, security consists of "…reducing the risk to critical infrastructure by physical means or defensive cyber measures to intrusions, attacks or the effects of natural or manmade disasters…,"[9] and resilience is the ability to "...withstand and recover rapidly from disruptions...deliberate attacks, accidents or naturally occurring threats or incidents."[10]

Taken together, these two concepts embody the pragmatic approach that the DHS employs to fulfill its mission. Unlike other national security agencies, the DHS consciously employs a risk-based methodology. It recognizes the breadth of its mission set makes the provision of coverage for every possible threat to American security impossible. Therefore, the DHS seeks to focus on threats with the potential greatest impact and encourages preventive and protective activities at the level best suited to reduce those threats. As a result, the DHS mission set is predominantly defensive, focuses more on domestic threats, and includes many law-enforcement authorities. The DoD and DHS mission sets complement each other as threats span geographic boundaries and continually evolve. Thus, the DHS addresses the most dangerous threats comprehensively while consciously accepting risk in areas it deems less consequential to American security.

One complementary aspect of the DHS mission is to guard the nation from hazards—short of war—that imperil the homeland. These threats are multifaceted, ranging from information and cyber warfare as well as election interference to terrorism and human and drug trafficking.[11] As these types of attacks become an increasingly common tool of American adversaries, the capabilities of the DHS to combat it are increasingly critical.[12] As a result, the DHS and DoD missions are not only complementary but also overlap considerably. As retired Colonel Tom Goss explains:

> ...the U.S. is confronted with a spectrum of threats ranging from traditional national security threats (e.g., ballistic missile attack) to law enforcement threats (e.g., drug smuggling). Conceptually, this threat-environment mosaic is not a clear matrix ... but rather a threat "spectrum"—a range of hostile challenges from what Americans consider "war" to what most label as "crime." ... This is a conceptual spectrum with clear definitions at both ends and less clarity in the middle, where the two ends blend together.[13]

Thus, the distinction between the two departments is increasingly blurred with each passing year. A concept of national security solely focused on military threats and solutions is incomplete and insufficient. A DoD-centric view of security reinforces patterns of investment into little-used capabilities and fails to recognize hazards of greater frequency and cumulative impact. For these threats, the capabilities that the DHS offers to the national security enterprise are vital.

DEVELOPING STRATEGY TO ALLOCATE RESOURCES: ALL HAZARDS, RISK AND RESILIENCE

The DHS risk-based approach to resourcing considers "all hazards"—or every relevant threat—to develop its strategy. It executes its mission through cooperation along three broad lines of effort:[14]

1. First is provisioning national-level capabilities such as border, maritime, and transportation security. This accounts for roughly 61% of DHS 2020 budget.[15]

2. Second is the regulation of direct security and resilience investments. While this is not a direct budget expenditure, one study estimated the cost to the US economy in 2017 of DHS-directed regulation was $57.2 billion, a sum nearly equal to the entire DHS budget for that year.[16]

3. Third, the DHS works through public-private partnerships to align private-industry incentives to security priorities and provide disaster response. For example, the Cybersecurity and Infrastructure Security Agency (CISA) provides oversight of critical infrastructure security and resilience investment. This is largely carried out by the private sector as well as state, local, and tribal governments. The Federal Emergency Management Agency (FEMA) aids in those instances where a disaster overwhelms the response capacity of the private sector and local government. These activities account for roughly 34% of DHS budget expenditures.[17]

This "all hazards" orientation attempts to prioritize across the range of potential risks and avoid excessive focus on any one threat at the expense of another.[18] This is a fine balance, the success of which can only be assessed post hoc. The DHS attempts risk mitigation in multiple ways: through interdiction of threat actors, through hardening or defense of vulnerabilities, and by building individual and private-sector resilience consistent with privacy, transparency, civil rights, and civil liberties.[19]

DHS History and Organization

The 9/11 attacks graphically demonstrated the limits of geographic isolation, which had underpinned a sense of relative American invulnerability dating from the eighteenth century. Although the September 11 attacks were the catalyst for governmental reorganization, cracks in the bureaucratic foundation had been evident for decades. From the increasing devastation of natural disasters such as Hurricane Andrew in 1992, to the bombing of the World Trade Center in 1993, and concern over the security of information systems as Y2K approached, the US faced mounting challenges to the security of its information, political, and economic systems. This vulnerability was laid bare on 9/11, along with the urgent need for a new approach to counter threats to the homeland.[20]

The Homeland Security Act of 2002, like the National Security Act of 1947, was a sweeping and comprehensive reorganization of American government. Broader than its 1947 counterpart, it addresses lateral coordination among federal departments, vertical coordination among levels of government, and interaction with the private sector. In the nearly two decades since the creation of the DHS, its mission has steadily expanded.

Meeting this growing mandate has proved challenging.[21] The DHS lacks unifying legislation akin to the 1986 Goldwater-Nichols Act that built jointness within the DoD.[22] Currently, there are few incentives for members of individual departments under the DHS umbrella to "see the elephant." Each agency has its own mission and mandate. As a result, priorities are often more parochial than holistic. This in turn increases the overall risk of critical mission failure.

To operate effectively, the DHS must organize the resources and efforts of entities over which it has no direct control. To accomplish this, the department employs a hybrid approach to strategy and resourcing. Certain DHS missions involve deployment of forces to directly defend against threats and mitigate risk. Examples include the US Coast Guard (USCG) and Customs and Border Patrol (CBP) with law-enforcement functions to deter and apprehend threat actors consistent with rule of law.[23] The DHS accomplishes other portions of its mission through coordination and partnerships, as in the case of critical infrastructure protection and disaster response.

The nation's federal structure and private-sector orientation compel the employment of a decentralized approach to mission accomplishment. For instance, FEMA reinforces local governments during incident response, mitigation, and recovery against recurrent threats, such as severe weather, for which there is no deterrent.[24] The CISA, created in 2018, oversees another such decentralized process; protecting digital networks and critical infrastructure systems. The DHS functions as the Lead Federal Agency (LFA), meaning that it coordinates, rather than directs, other departments and agencies through the CISA.[25] The mission requires the DHS to harmonize security across sixteen separate infrastructure sectors, which impact all aspects of US economic activity.[26]

While the CISA functions as the LFA, other federal departments and agencies—acting as Sector Specific Agencies (SSAs)—establish and manage the planning forums of state, local, and tribal governments, along with key private-sector organizations that control critical infrastructure. These groups conduct risk assessments, determine appropriate invest-ment levels, and coordinate response planning to protect infrastructure systems.[27] Thus, although the DHS has the lead role, its actual ability to direct operations is limited. For example, the Environmental Protection

Agency (EPA) functions as the SSA for the nation's water supply under the aegis of the DHS as the LFA. In this capacity, it is the EPA who works with water utilities and governments to oversee risk assessment and strategy development.[28]

Planning and investment forums like the one described fall under the broad label of Public-Private Partnerships (PPP). In these cases, the DHS has limited authority to prioritize efforts or produce an integrated cross-sector resourcing strategy. In practice, SSAs tend to focus on current challenges rather than emergent risks. The result is short-term thinking that leaves agencies preoccupied with costly consequence management efforts rather than focused on mitigation efforts that could avoid catastrophes altogether or investments that could have lowered the cost of the resulting consequence management efforts.[29] The current process hinders holistic thinking and integration of strategies, priorities, programs, and budget. This grim picture is further complicated by congressional oversight that spans multiple jurisdictions and authorizing committees, which inevitably produce conflicting priorities.

The DHS also liaises directly with foreign governments to identify and mitigate security threats originating abroad through a presence on certain Department of State Embassy Country Teams.[30] The DHS also engages directly with private-sector entities, such as businesses, non-profits, and academia to provide incentives for voluntary efforts to increase security, resilience, and continuity of operations in the event of an attack or natural disaster. During his swearing in ceremony, the first DHS Secretary, Tom Ridge observed:

> We must open lines of communication and support like never before between agencies and departments, between federal and state and local entities and between the public and private sectors. We must be task oriented. The only turf we should be worried about protecting is the turf we stand on.[31]

The DHS, like other national security agencies, must anticipate diverse threats, define security outcomes, and prioritize efforts toward the greatest threats, while accepting risk in other areas. However, DHS missions are often ambiguous. They tend to involve continual operations among civil society and encompass persistent, difficult-to-address threats, such as cyber disruption or violent extremism. The benefits of investment are difficult to assess, as they resemble insurance market outcomes—loss avoidance—and thus are difficult to weigh and justify against costs. Further, the DHS strategy development and resource alignment processes are less well-developed than other federal departments. As a result, they tend to be driven by subcomponent (e.g., individual agency) requirements rather than national priorities.[32]

The breadth of coordination required to fulfill this mandate is enormous. It is also complicated by the fact that the security priorities of these independent actors do not always align with the DHS mission. Challenging things further is the reality that the necessary expertise is dispersed across the federal government, local governments, and the private sector. Given that the DHS budget is insufficient to comprehensively resource against every threat type, it becomes obvious why the DHS has no option but to assume a risk-based approach to strategy and resourcing. Yet despite these shortcomings, the broad reach and convening authority of the DHS remains one of its most valuable resources.

Currently, the DHS is actively engaged on many fronts: combatting terrorism, violent extremism, intellectual property theft, election tampering, emergency management as well as customs and border patrol. This diverse list includes threats with radically different risk profiles, which in turn require different solutions and resources.[33] The DHS urgently needs clear oversight lines to improve and rationalize its focus of effort.

RESOURCING STRUCTURE AND TRENDS

A secure homeland requires coordinated investment across all levels of government and the private sector. As explained, many relevant stakeholders and budgets lie outside of the control of the DHS, which in turn requires multiple methods to resource and integrate its operations. The DHS directly funds its federal missions, employs regulation or federal contracting standards to direct private-sector activity, and uses incentives to induce voluntary investment. It accomplishes all this by influencing the actions of parties outside its control and by developing and providing response plans to concentrate capabilities during incidents.[34] The approach yields mixed results.

The budget and mission of the DHS have grown erratically since its inception. Prior to its establishment, the US government budgeted $9 billion for homeland security functions in 1995, $16 billion in 2001, and $19.5 billion in 2002 in the immediate aftermath of the 9/11 attacks. Later in 2002, Congress added an additional $9.8 billion in emergency response funding.[35] In 2003, Congress appropriated $37.7 billion for the DHS's first department budget. From 2004 onward, the budgets have consisted of yearly base funding with additional resources from the Disaster Relief Fund (DRF) allocated during emergencies. DHS budgets have increased at an average real rate of 3.6 percent in the two decades since 2002.[36] Since its founding, DHS funding has been late, variable, and bundled with other federal departments. The DHS has not had an authorization bill in place at the start of its fiscal year since 2010.[37] Congress apportions the DRF by disaster rather than by program. As a result, the DRF sometimes comprises a large budget share, as in 2018 when the DRF accounted for 43% of the DHS budget.[38] The erratic pattern of disaster relief funding makes consistent programming difficult.[39]

Nonetheless, DHS budget allocations across other mission sets have been relatively consistent. In FY 2003, budget priorities were border security (28%), biological terrorism (16%), aviation security (13%), first-responder support (9%), and sharing information/technology protection (2%). The budget also included DoD homeland security initiatives (18%) and other non-DoD homeland security activity (14%).[40] In 2021 the DHS chose to distribute budget shares by agencies rather than by function, but with little change to overall allocations.

Despite this budgetary consistency, the threats have multiplied since the DHS's founding. There has also been an increased emphasis on physical security and antiterrorism. The DHS mission has also expanded to encompass areas such as cyber incidents and election interference. However, resource allocations have remained the same. Thus, the DHS needs better alignment of resources to meet national risk-mitigation priorities.

Direct Spending to Establish Inherently Federal Capabilities

Congress funds the DHS's base budget from the discretionary portion of the federal budget. These funds support forecasted operations and comprise, on average, 78% of the department's yearly appropriations. DRF appropriations, spent through FEMA, comprise the remaining 22% of DHS average appropriations. [41] As previously noted, DRF funding varies widely from year to year, which makes programming difficult.[42]

The DHS base budget supports national-level functions such as the customs and border patrol as well the Coast Guard.[43] In addition, it funds grants to state, local, and tribal governments to support security and resilience investments. The DHS budget also funds permanent staff which enable its Lead Federal Agency and public-private-sector coordination missions. For example, FEMA staffs a standing structure for disaster response. In the event of an incident, FEMA receives DRF funding and deploys response teams that are augmented with expertise from across the federal government to assist state and local governments.[44] This is authorized through the 1988 Stafford Disaster Relief and Emergency Assistance Act, which outlines the criteria and processes for state requests and deployment of federal assistance.[45]

Like FEMA, the CISA provides cyber disaster prevention and response expertise to industry and government.[46] Jointly, FEMA and the CISA are tasked to secure the nation's economic and political systems from disruption, and yet their base funding of $4.2 and $3.2 billion respectively comprises only 8% of 2020 DHS budget.[47] This would seem an inadequate amount to protect the nation's vital governance and economic functions.

The DHS also provides grants as a means to align state and local security and resilience investment with national priorities.[48] From 2003–2016, DHS awarded over $33 billion in preparedness grants for investment in protection, mitigation, and recovery from terrorist attacks, major disasters, and other emergencies.[49] The grants are intended to address high-risk systems, such as transit lines and port structures, which are expensive for local governments to secure themselves. However, grant funding has been somewhat erratic. With peaks of $3.53 billion in 2004 and $3.49 billion in 2009, the sustained average has been $1.5 billion across fourteen programs.[50] The award system has been criticized in audits by the Government Accountability Office and others for insufficient focus, unclear outcome criteria, and a need for peer review.[51] There is consensus that grant funds do not always go toward high-risk areas or address emergent risk. For instance, there are few grants intended to improve the ability to counter nontraditional and emergent threats such as domestic terrorism, pandemic response, and new forms of cyberattack.[52]

The DHS also assesses fees for the provision of security where the service directly benefits the user, and the department can efficiently collect revenue. Fees such as CBP inspection charges on commercial carriers offset service costs and reflect the true price of asset operation.[53] The DHS also applies this practice to individuals who directly benefit from services. This includes such items as applications for naturalization, transportation-security cards, and other travel fees. In 2016, the DHS collected over $15 billion in user fees, with revenues being prioritized to extending services and expanding expenditures to counter high-risk threats.[54]

Directive Measures: Regulation and Contracting

The DHS also regulates risk-producing activities outside its control and leverages the expertise of owner/operators to ensure security measures and controls. The development of Chemical Facility Anti-Terrorism (CFATS) standards is an example of this. Beginning in 2007, the DHS published risk-based performance standards for chemical processing, with the actual implementation of security programs left to owner operators and then verified by DHS inspectors.[55] Through this process, the DHS regulation leverages industry expertise to develop solutions and find efficiencies.

The purchases and financial flows of the US government also provide the DHS with the opportunity to shape private-sector security. Federal purchases of goods and services total more than $600 billion yearly, and the sum total of US government expenditure and transfers in 2020 exceeded $4.1 trillion dollars.[56] All these transactions flow through the financial system following procedures governed by federal regulation.

The Department of Defense Cybersecurity Maturity Model Certification (CMMC) leverages this relationship. The DoD requires companies that bid on defense contracts to certify compliance with cybersecurity standards through external audits of their policies and procedures.[57] The CMMC protection of the defense industrial base secures critical infrastructure vital to national and economic security. As these cybersecurity standards gain wider use, they will become the default protection standard for industry.

Indirect Approaches

The DHS uses public-private partnerships, such as those outlined in its National Infrastructure Protection Plan (NIPP) to incentivize industry and individual investment in security and resilience.[58] Public-private partnerships employ a voluntary approach through which the DHS reduces the cost of security investment by providing assessments, intelligence, and technology free or at greatly reduced costs so that companies can direct their investment toward security and resilience improvements.[59] For instance, the National Institute of Standards and Technology (NIST) developed the 800-171 cybersecurity standards as an easily adaptable framework for private-sector firms to build their own cyber-protection programs. The government invests the time-intensive research and development so that industry can focus on protection. The DHS also offers incentives to encourage investment. The SAFETY Act of 2002 incentivized private-sector firms to develop technology to reduce risk in exchange for receiving limited-liability protection and a certified status for putting it into use.[60]

The DHS also extends incentives to encourage individuals and private-sector entities, as well as local and state governments to identify and address risks at their respective levels. For instance, the DHS can conduct mass traveler screening at ports of entry and departure themselves at great cost, or it can employ a tiered approach in which travelers self-select into lower-risk groups and consent to background checks. The trusted traveler program by the Transportation Security Administration (TSA) employs this technique, whereby individuals pre-screen onto a preapproved list. More thorough scrutiny can then be applied to unenrolled populations. Though in some cases, this does take risk monitoring out of the direct control of the DHS, the department extends its resources by leveraging the activities of individuals and the private sector.

Over time, the DHS has refined its partnership framework to better define priorities. An example of this is the designation of lifeline sectors in the 2013 National Infrastructure Protection Plan (NIPP). Lifeline sectors are those in which disruptions carry a higher risk of death, injury, or economic loss. An extension of this concept is the designation of Strategically Important Critical Infrastructure (SICI).[61] This allows for greater information sharing and better response toward systems and assets that deliver vital functions such as electric power or communications.[62] To facilitate protection and support for these sectors, the DHS has championed the creation of state-level and private-sector information-sharing nodes.[63] However, these structures receive minimal public funding, which limits the extent of their reach into the private sector.[64]

Finally, the DHS facilitates a variety of exercises to build relationships, enable information exchange, risk assessment, planning, and preparation. The "GridEx" series of exercises, sponsored by the North American Electric Reliability Corporation and supported by DHS, serves as an example for coordination across government and private industry.[65] These exercises test and refine response systems that are valuable in the event of disaster. They are low-cost, high-return events that build relationships with individuals and the private sector.

CONCLUSION

The DHS is entering its third decade of existence. To quote Winston Churchill, "Now this is not the end. It is not even the beginning of the end. But it is, perhaps, the end of the beginning."[66] Homeland security missions are here to stay, and they are increasingly complex and important. The threats they combat are as dangerous as the military threats currently facing the US. Yet the DHS missions are perennially underresourced. Further, their existing risk assessment and resource allocation systems require improvement.[67]

It has been two decades since 9/11. It does not take much analysis to establish that the threats facing the nation have changed. Since the turn of the century the country has experienced terrorism, increasing cyberattacks and intellectual property theft, an increasing level of natural disasters exacerbated by climate change, and a global pandemic. These are all DHS mission sets. The DHS is neither a peripheral nor a temporary addition to US national security. However, it is resourced and organized as though it were both. To elevate the DHS to an appropriate level and provide Americans with the full range of security capabilities, we make the following recommendations:

1. The DHS requires a presence in national security decision-making on par with the gravity of its mission. This requires representation in the National Security Council and presence in the appropriate policy development and decision-making bodies.[68]

2. As DHS missions have expanded and evolved, the department requires regular and consistent legislative authority that refines its mission and provides authority commensurate with its responsibilities. The DHS has not received a reauthorization of its mission since its inception in 2002.[69] By contrast, the Department of Defense receives an annual authorization bill, which clarifies policy and focuses effort.

3. DHS reauthorization should come with increased authority to direct interagency efforts and engage the private sector to better understand risk, measure effectiveness, and allocate resources toward the most effective mitigation efforts.

4. The DHS requires a consolidation of its congressional oversight as well as authorization and appropriation relationships. Currently, the DHS is overseen by up to 100 subcommittees across both houses of Congress.[70] By contrast, the Department of Defense and Department of State each report to four committees with subordinate subcommittees.[71] These multiple lines of jurisdiction lead to confusion, misallocation of resources, and gridlock. The department should have no more than one authorizing and one appropriating committee in each house of Congress commensurate with other federal departments.[72]

5. The DHS should receive funding that is predictable, stable, and flexible to meet the department's missions.

The Department of Homeland Security has now operated as a single entity for nearly two decades. Yet, it is plagued by internal weaknesses that rob the department of unified strategy development and a clear understanding of the risks, impacts, and true costs of mitigation. The DHS would improve its efficiency with greater centralization as well as the ability to shift personnel and resources within the department.[73]

The DHS also requires improvements in its strategy development and resource allocation processes. It conducted its last Quadrennial Strategy Review in 2014. This process should resume to assess extant and emergent risk, clarify priorities, rigorously develop strategy, and guide and give order to resource allocation. Documentation from this process can then guide strategic planning, programming, and budgeting, thereby lessening the large priority swings that accompany administration changes. The extensive investment into border barrier systems conducted by the Trump administration and then suspended by the incoming Biden administration is instructive in this regard.

Among the federal departments, the DHS has unique convening authority that enables coordination across government and the private sector.[74] Public-private relationships are vital to build trust that allows for coordination, assessment, planning, and response. The DHS also has authority to conduct confidential dialogue with the private sector. These efforts should mature and expand to raise the volume and quality of communication between the public and private sector.[75] Information sharing between the public and private sector requires structures with expertise and equipment to facilitate information flow in a robust and timely manner. To facilitate information sharing, the DHS should request additional resources to support information-sharing structures and increase engagement with industry.

The DHS mission has expanded as technology has diminished the value of geographic isolation. The mission set of the DHS will continue to grow as competitor nations become more aggressive, natural hazards grow in severity, and technology enables more adversaries to reach the US homeland. Nonmilitary threats will increase. The nation needs a holistic approach to its national security with the DHS in the lead to counter nonmilitary threats.

NOTES

1. Thomas Ridge, "Swearing in Remarks," *Washington Post*, October 8, 2001, https://www.washingtonpost.com/wp-srv/nation/specials/attacked/transcripts/bushtext_100801.html.
2. I wish to recognize with gratitude the advice and research guidance provided by the Honorable Caitlin Durkovich, former Assistant Secretary of Homeland Security, in the formulation of this chapter.
3. U.S. Department of Homeland Security, "Creation of the Department of Homeland Security."
4. Morgan, *The Impact of 9/11 on Politics and War*, and U.S. Department of Homeland Security, "DHS Mission."
5. Warrick and Durkovich, *Future of DHS Project*, 1.
6. Wolf, *FY 2021 Budget in Brief*, 1.
7. Ibid., 2.
8. Warrick and Durkovich, *Future of DHS Project*, 1.
9. Barack Obama, "Directive on Critical Infrastructure Security and Resilience, Presidential Policy Directive 21," Washington, DC: White House, February 12, 2013), 12, https://www.hsdl.org/?abstract&did=731087.
10. Ibid.
11. Kinetic capabilities focus on destroying the enemy with lethal force, while non-kinetic methods can refer to cyber warfare, gray-zone activities, and criminal activity aimed at a government, information warfare, psychological operations, among others.
12. The term "gray zone" refers to the employment of the instruments of power to achieve national objectives that falls short of physical conflict. Gray-zone methods may exploit ambiguity in international law, an inability to attribute actions, or the opponent's decision that the activity does not merit a response.
13. Goss, "Who's In Charge," 1.
14. The "all hazards" approach considers naturally occurring threats such as weather hazards, technological threats such as aging and breakdown risks, and human-based hazards such as cyber hacking or terrorism. See Federal Emergency Management Agency, *National Preparedness Goal*, 3–4.

15. Kirstjen M. Nielsen, "FY 2020 Budget in Brief" (Washington, DC: Department of Homeland Security, November 8, 2019), 7, https://www.dhs.gov/publication/fy-2020-budget-brief.

16. Crews, *Tip of the Costberg*, 72–77; and Painter, *Trends in the Timing and Size of DHS Appropriations*, 5.

17. Kirstjen M. Nielsen, "FY 2020 DHS Budget in Brief," 7, https://www.dhs.gov/publication/fy-2020-budget-brief. The remaining 5% of the DHS 2020 budget is devoted to headquarters and other functions.

18. See "National Risk Management," Cybersecurity and Infrastructure Security Agency, https://www.cisa.gov/national-risk-management.

19. U.S. Department of Homeland Security, "Guiding Principles."

20. For an in-depth discussion, see Thomas H. Kean, et. al., *The 9/11 Commission Report* (Washington DC: National Commission on Terrorist Attacks, July 22, 2004), https://www.9-11commission.gov/report/911 Report.pdf

21. Warrick and Durkovich, *Future of DHS Project*, 29.

22. Ibid., 30.

23. "Mission of the Customs and Border Patrol," Customs and Border Patrol, U.S. Department of Homeland Security, last updated December 18, 2020, https://www.cbp.gov/about; and "Mission of the U.S. Coast Guard," U.S. Coast Guard, U.S. Department of Homeland Security, https://www.uscg.mil/About/Missions/.

24. "Mission of the Federal Emergency Management Agency," Federal Emergency Management Agency, U.S. Department of Homeland Security, last updated January 4, 2021, https://www.fema.gov/about/mission.

25. 43 CFR § 46.220 - How to designate lead agencies. | CFR | US Law | LII / Legal Information Institute, https://www.law.cornell.edu/cfr/text/43/46.220

26. CISA identifies the sixteen critical infrastructure sectors; see https://www.cisa.gov/critical-infrastructure-sectors.

27. Cybersecurity and Infrastructure Security Agency, *NIPP 2013: Partnering for Critical Infrastructure Security and Resilience*, 15–16 (Washington DC: Department of Homeland Security, 2013), https://www.cisa.gov/sites/default/files/publications/national-infrastructure-protection-plan-2013-508.pdf.

28. Ibid., 10–12.

29. Warrick and Durkovich, *Future of DHS Project*, 16–17.

30. Dorman, *Inside a U.S. Embassy*, 69.

31. Thomas Ridge, Swearing in Remarks," *Washington Post,* October 8, 2001, https://www.washingtonpost.com/wp-srv/nation/specials/attacked/transcripts/bushtext_100801.html.

32. See Warrick and Durkovich, *Future of DHS Project,* 29; and U.S. Department of Homeland Security, "Joint Requirements Council," last updated April 15, 2019, https://www.dhs.gov/joint-requirements-council.

33. For an in-depth view of this concept, see chapter 2 of Ted G. Lewis, *Critical Infrastructure Protection in Homeland Security.*

34. Warrick and Durkovich, *Future of DHS Project,* 6–9.

35. The DHS FY 2003 budget provides a funding history, see George W Bush, "Securing the Homeland, Strengthening the Nation," 8 (Washington, DC: Department of Homeland Security, 2003), https://www.dhs.gov/publication/securing-homeland-strengthening-nation.

36. Painter, *Trends in the Timing and Size of DHS Appropriations,* 5

37. Ibid., 1.

38. Ibid., 5.

39. Ibid., 1–2.

40. George W. Bush, "Securing the Homeland, Strengthening the Nation," (Washington, DC: Department of Homeland Security, 2003), 8, https://www.dhs.gov/publication/securing-homeland-strengthening-nation.

41. Author calculations based on data from Painter, *Trends in the Timing of DHS Appropriations,* 5.

42. Ibid., 4.

43. "TSA Mission," Transportation Security Agency, February 15, 2021, https://www.tsa.gov/about/tsa-mission; and "ICE's Mission," Immigration and Customs Enforcement, https://www.ice.gov/mission.

44. "FEMA Field Operations," Federal Emergency Management Agency, U.S. Department of Homeland Security, last updated November 17, 2020, https://www.fema.gov/about/offices/field-operations.

45. Bruce R. Lindsay, *Stafford Act Declarations 1953-2016: Trends, Analyses, and Implications for Congress* (Washington, DC: Congressional Research Service, 28 August 2017), https://crsreports.Congress.gov/product/pdf/R/R42702.

46. "ICS-CERT Overview," Cyber and Infrastructure Security Agency, U.S. Department of Homeland Security, https://us-cert.cisa.gov/ics.

47. Kirstjen M. Nielsen, "DHS FY 2020 Budget in Brief," 7, 45, 51, https://www.dhs.gov/publication/fy-2020-budget-brief.

48. See "National Preparedness Goals," Federal Emergency Management Agency, U.S. Department of Homeland Security, https://www.fema.gov/emergency-managers/national-preparedness/goal, and mission areas of prevention, protection, mitigation, response, and recovery.

49. Shawn Reese, *Department of Homeland Security Preparedness Grants: A Summary and Issues*, (Washington, DC: Congressional Research Service, October 28, 2016) 13–15, https://crsreports.Congress.gov/product/pdf/R/R44669.

50. Ibid., 5.

51. U.S. Government Accountability Office, *Homeland Security Grant Program*; and U.S. Government Accountability Office, *Homeland Security*.

52. Reese, *Department of Homeland Security Preparedness Grants*, 5–8.

53. "Fee Schedule," Customs and Border Patrol, , U.S. Department of Homeland Security, https://www.cbp.gov/trade/basic-import-export/user-fee-table.

54. U.S. Government Accountability Office, *DHS Management*.

55. Frank Gottron, "Chemical Facility Anti-Terrorism Standards" (Washington, DC: Congressional Research Service, February 14, 2020), https://crsreports.congress.gov/product/pdf/IF/IF10853; and Title 6 CFR Part 27, reauthorized 22 July 2020, https://www.ecfr.gov/current/title-6/chapter-I/part-27; and Bridget Johnson, "Three-Year CFATS Reauthorization Signed Just Before DHS Program Was Set to Expire," *Homeland Security Today*, July 22, 2020, https://www.hstoday.us/subject-matter-areas/emergency-preparedness/three-year-cfats-reauthorization-signed-just-before-dhs-program-was-set-to-expire/
.

56. Courtney Bublé, "Federal Contract Spending Reaches Its Highest Level Ever in Fiscal 2019, Marking 4 Straight Years of Growth," *Government Executive*, June 26, 2020, https://www.govexec.com/management/2020/06/federal-contract-spending-reaches-its-highest-level-ever-fiscal-2019-marking-4-straight-years-growth/166484/.

57. "Cybersecurity Maturity Model Certification Overview," Undersecretary for Acquisition and Sustainment, U.S. Department of Defense, https://www.acq.osd.mil/cmmc/.

58. Cybersecurity and Infrastructure Security Agency, *NIPP 2013: Partnering for Critical Infrastructure Security and Resilience*, 7 (Washington DC: Department of Homeland Security, 2013), https://www.cisa.gov/sites/default/files/publications/national-infrastructure-protection-plan-2013-508.pdf.

59. "CMMC Frequently Asked Questions," Acquisition and Sustainment, Office of the Under Secretary of Defense, https://www.acq.osd.mil/cmmc/faq.html.

60. U.S. Department of Homeland Security, "DHS SAFETY Act."

61. Cybersecurity and Infrastructure Protection Agency, National Infrastructure Protection Plan, (Washington, DC: Department of Homeland Security, 2013), 16–17.

62. Angus King and Michael Gallagher, "Cyberspace Solarium Commission Report," (Washington, DC, March 2020), 6, https://www.solarium.gov/report.

63. Cybersecurity and Infrastructure Protection Agency, "Critical Infrastructure Threat Information Sharing Framework," (Washington, DC: Department of Homeland Security, October 2016), 26–27, https://www.cisa.gov/sites/default/files/publications/ci-threat-information-sharing-framework-508.pdf.

64. Weiss, Eric N., "Cybersecurity and Information Sharing," (Washington, DC: Congressional Research Service, 29 April 2015), https://crsreports.Congress.gov/product/pdf/IF/IF10163.

65. Ciampoli, "Grid Exercise Evolves as Public Power Participation Increases."

66. Attributed to Winston Churchill from remarks at the Lord Mayor's luncheon at Mansion House, Edinburgh, Scotland in October 1942, retrieved from https://winstonchurchill.org/the-life-of-churchill/war-leader/1940-1942/autumn-1942-age-68/.

67. Warrick and Durkovich, Future of DHS Project, 5–6; and King and Gallagher, "Cyberspace Solarium Commission Report," 1–2.

68. Warrick and Durkovich, Future of DHS Project, 11, 17.

69. Warrick and Massa, The Future of DHS Project , 19.

70. Rosenzweig, "Streamlining Congressional Oversight of DHS."

71. Warrick and Massa, The Future of DHS Project, 13.

72. Ibid., 5.

73. Warrick and Durkovich, Future of DHS Project, 30.

74. Ibid., 2.

75. King and Gallagher, "Cyberspace Solarium Commission Report," 96–109.

CHAPTER 12

CONCLUSION

Susan Bryant

God grant me the serenity to change the things I cannot change ... The courage to change the things I can, And the wisdom to know the difference.

—Reinhold Neibuhr[1]

INTRODUCTION

At the height of the Great Depression, American theologian Reinhold Niebuhr wrote "The Serenity Prayer."[2] Although it has been edited and adapted by various churches and organizations, it provides a useful frame when considering the complexities associated with resourcing US national security.

Given the multiple legally enshrined and interlocking processes associated with strategic resourcing, there are many items which an individual or an organization simply cannot change, regardless of their knowledge of the process or their position within the system. Further, the points at which policy makers and politicians can effect change require a thorough understanding of the system, which in turn requires considerable knowledge—if not necessarily wisdom. And finally, although the bureaucratic wrangling necessary to effect systemic change does not require battlefield courage to overcome the threat to life and limb, anyone who has undertaken an effort to change the ways in which a large bureaucracy works understands that at minimum, persistence and a large emotional reserve are required. Therefore, Niebuhr's divinely sought combination of serenity, courage, and wisdom represents an ideal set of characteristics for all strategists and policy makers, whether aspiring or practicing.

This conclusion will also adapt Niebuhr's categories of "things that cannot change" versus "things that can" as it attempts to analyze and amalgamate the major findings in this book. It will begin with the immovable objects and immutable facts that confront anyone trying to coherently resource US national security, regardless of the office or position they hold within the system. It will then consider those items, which—with a combination of courage, knowledge, opportune timing, and luck—skillful policy makers and strategists can hope to change. That said, we recognize that although change is possible, it will not be easily achieved.

THE "UNCHANGEABLE THINGS"

1. The US Constitution

The first unalterable object is the US Constitution itself. The Founders carefully designed the government with divided powers to create an intentional inefficiency.[3] Thus, their finely crafted system of checks and balances is not supposed to work rapidly or cost effectively. As a result, one of the principal obstacles to the rationalization of the ends, ways, and means of any American strategy is the foundation of the system itself.

As Tom McNaugher points out in chapter 7 of this book, within the US government, the people who appropriate money for national security are entirely different from the people who request and spend it.[4] Therefore, the process will never be rational, so to speak, but rather will be an outcome created by the political machinations of those legions of people who participate in the system and who will remain largely anonymous as they go about their daily work. This is a fundamental reality that every strategist, politician, and policy maker must accept and work around.

2. Enduring Security Priorities

Although not legally enshrined, there has been a remarkable consistency in how the US government and, by extension, the citizens who elected it view their national security priorities. Since it was first published during the Reagan administration, every US National Security Strategy, regardless of the political party in power at the time, has consistently identified similar overarching priorities. They are:

- Security: This subsumes the traditional notion of national defense, or an ability to deter or defeat threats before they reach US territory.
- Economic Prosperity: The element is both an enabler of capabilities and an end in itself. Consistent growth in living standards promotes stability. Serious threats to economic prosperity, whether human, technological, or natural in origin, are potential threats to national security.
- Projecting American Influence: US national security, especially in the post–World War II era has stressed the engagement of nations and organizations to shape an international system favorable to US interests.
- Order: The protection of US territory to include its people and the political and economic features that comprise the American way of life. These span individual freedom, representative government, and free markets.[5]

These principles are enshrined in the nation's founding documents such as the preamble of the US Constitution, which states the document was established to "ensure domestic tranquility, provide for the common defense, promote the general welfare, and secure the blessings of liberty."[6] Therefore, although the printed National Security Strategy only dates from the late twentieth century, the sentiments underpinning them all date back to the late eighteenth century.

3. The Budget Math

In chapter 2 of this book, Mark Troutman observes that over the past two decades, when faced with a choice between the enduring priorities of national security and economic prosperity, the American people have shown an increasing willingness to prioritize prosperity. This is a particularly salient observation in light of the worsening budget crisis currently confronting the United States. The US is on an unsustainable resourcing trajectory as mandatory spending continues to crowd out all other priorities, meaning that unless there are substantial changes to how the American government chooses to tax and spend, prosperity and security are in direct competition for a decreasing share of the budget. Strategically, the American government's current alignment of ends to means—of revenue generation to government spending—is incoherent and is running a high risk of failure if significant reform is not immediately undertaken.

A fundamental strength of the United States throughout history has been the vibrancy and resiliency of its economy. Thus, the increasing levels of deficit spending are undermining and imperiling one of the bedrock elements of American national power. To date, there has been no national will to take on needed reforms to restore balance and coherence to the system. The current bipartisan political tactic of "whistling past the graveyard" will inevitably have profound negative impacts on the United States' ability to achieve its enduring security priorities.

4. The Bureaucracy

This volume has attempted to provide a coherent narrative about the processes by which the national security priorities of the United States are resourced. Despite the individual authors' obvious expertise and the editors' best attempts, coherence has proven challenging because the bureaucratic processes by which the United States resources its national security priorities are complex, inefficient, opaque, and often redundant. These processes are also subject to political agendas, partisan rancor, blatant disinformation, and the protection of agency equities over national interest. Thus, they have their own logic, which is nearly impenetrable to all but those who have spent years working within the system. As Robert Komer observed more than fifty years ago in *Bureaucracy Does Its Thing*:

> [Vietnam] shows the difficulty of changing institutional patterns even in the light of frustrating experience. Yet, large hierarchically organized institutions seem to be a fixed feature of the contemporary scene, indispensable to meeting many complex needs of society. Since we can't do without them, we have to pay the price of their built-in limitations to the extent that they cannot be altered. Thus, wise policy must take adequately into account the institutional realities that will largely shape its execution.[7]

Although the social fabric of the United States and the geopolitical realities it faces have evolved significantly in the ensuing half century since Komer penned these words, their bleak logic persists. As such, the bureaucratic system for resourcing national security cannot be "fixed" per se. Rather, it must be understood for what it is, and those working within it should seek to comprehend the "art of the possible"—those points within the processes where new ideas can best be injected and have the greatest chance of being funded and brought to fruition.

THE THINGS THAT CAN BE CHANGED

Although "fixing" the entire system is not possible, change is. It has occurred in the past and can happen again. The leaders of the United States government emerged from World War II recognizing that the completely divided structure of the military and security apparatus did not work well for an emerging superpower with global responsibilities.

The result of a combination of courage, wisdom, and sometimes vicious political wrangling, the National Security Act of 1947 was passed. It established a national security enterprise with the creation of the National Security Council, a National Military Establishment (that later became the DoD), and the Central Intelligence Agency (CIA).[8] Although there were a series of interim reforms in the ensuing decades since its passage, the most notable of which were the Goldwater-Nichols Act of 1986 and the Homeland Security Act of 2002, the system has persisted relatively unchanged for more than seven decades.

   Each author in this volume has an appreciation for the immutability of certain aspects of the resourcing process as well as a keen understanding of their complexities. Nonetheless, the majority of the authors do see points in the system where current processes could and should be changed. Perhaps even more comforting is that although each author was focusing on different facets of the resourcing labyrinth, several of them point to the same critical nodes as logical and possible places for change. Four of these recommendations stood out as being both possible and necessary.

Recommendation 1: Create a Congressional Select Committee on National Security Affairs and Restore the Annual Foreign Policy Bills

The National Security Act of 1947 has generally stood the test of time, having been amended on those occasions when glaring gaps were noted and were allowed by legislative consensus. The current difficulty is that more than seventy years later, the somewhat but never entirely discrete mission sets among the various agencies charged with ensuring US national security have blended together in ways that exceed individual departmental and agency mandates and competent oversight by multiple, separate congressional committees. In short, the landscape of national security and their associated mission sets have evolved significantly while the bureaucratic structures overseeing them have failed to keep pace.

It should not be lost on the reader that despite multiple attempts to reconcile whole-of-government approaches to national security, resourcing administration priorities occurs largely within the confines of separate departments and agencies as a result of the congressional committee system. As Tom McNaugher points out in chapter 7, the federal budget is the result of a circular process the end result of which is that individual agency budgets are put together and forwarded on to Congress.[9] Wholesale integration of priorities among agencies does not happen. Further, in terms of resourcing, the congressional committee structure effectively walls off separate funding streams for programs that are not entirely separate. In this volume, Tom McNaugher, Heidi Demarest, and Geoffrey Odlum each highlight resourcing conundrums associated with the current budget process as it occurs in annually in Congress.

The budget process is solidly stovepiped into neat columns by agency or department, while the missions they fund are crosscutting—in multiple senses of the word. To begin, certain missions involve multiple departments and/or agencies, while others transect federal, state, and local levels of government. The combination of mission sets and responsible departments constitute a matrix, with the responsible agencies and departments represented by columns and the missions articulated as rows. The difficulty arises when missions cut across multiple columns and rows or become diagonal lines transecting multiple rows and columns. The result is not merely inefficient; it risks mission failure.

For example, the Department of Defense and the Department of State both have responsibilities associated with Foreign Military Sales and Security Cooperation, but the agencies' budgets are funded separately. Another example of this phenomenon is the significant overlap that occurs in responsibility for cybersecurity. The DHS, the DoD, and the intelligence community are major participants, with Department of Treasury, the Department of Justice, and the Department of Energy also having significant supporting pieces of this puzzle. As described in chapter 11, the homeland security mission further crosscuts the federal, state, and local levels of government, and it also touches upon the remits of entirely nongovernmental entities, such as businesses and academia. Trying to understand how all of these unique strategies and separate lines of funding intersect and interact is a monumental, if not futile, task.

In chapter 3, Heidi Demarest argues for the creation of a congressional "select committee on national security affairs" to provide oversight and responsibility of the joint space between Congress and the executive branch and thus improve Congress's ability to allocate resources for national security.[10] If created, this committee could create synergy between the House and Senate foreign relations committees, budget committees, armed services committees, intelligence committees, and appropriations committees and provide oversight that transcends the currently stovepiped system. Demarest further argues that to operate effectively, this select committee would require the authority to affect decisions about resourcing and contain enough resident expertise to deliberate in a substantive and productive way about aligning national means to strategic ends.[11]

In chapter 5, "Determining and Resourcing Diplomatic Priorities," Geoffrey Odlum also addresses the need for congressional reform as it pertains to coherently resourcing American foreign policy. One of his recommendations amounts to returning to formerly enshrined practices rather than the creation of something new, which would seem somewhat easier to achieve. He notes that "Congress's abandonment of the foreign affairs authorization process [in 1986] in favor of reliance on the appropriations process to approve foreign affairs funds and direct how they are spent at the country level" is imperiling the creation and oversight of a coherent American foreign policy altogether.[12] Odlum argues that the current processes have "allowed short-sighted politics and parochial horse-trading to shape the SFOPS budget."[13] He further opines that it is only through the restoration of the annual Foreign Assistance

Authorization Act in Congress that the US can hope to fund or executive a foreign policy that is capable of achieving its strategic objectives.

Although neither of these recommended reforms would be an easy sell in a Congress that is currently characterized by party line votes and vindictive partisan rancor, they are both necessary for increasing the coherence of American foreign policy and the execution of the president's stated national security goals. Maintaining the system in its current form severely reduces the chances that the United States will be able to either fund or implement a coherent national security strategy.

Recommendation 2: Establish an NSC-Managed Periodic Assessment of Strategic Resourcing

In chapter 4 of this book, Jason Galui briefly recounts the creation of the Scowcroft model of the National Security Council process. He observes where and how the executive branch can have the greatest impact on the resourcing decisions that animate America's national security priorities and perhaps ameliorate the strategic confusion from which he believes the United States is currently suffering. Specifically, he recommends that the National Security Council directors maintain a close relationship with their counterpart OMB examiners in order to ensure the greatest chance that the president's strategic priorities are actually resourced. [14]

In chapter 5, Geoffrey Odlum also points to the NSC process as a potential avenue for improving the overall coherence national security resourcing, arguing that the National Security Council should approach national security strategy-making more holistically when preparing the president's annual budget submission to Congress and set national security goals, objectives and resource needs in a more integrated and cost-effective fashion.[15]

To that end, the authors in this book would like to endorse the recommendations of the Atlantic Council's report, *Connecting Strategy and Resourcing in National Security*. Its authors, including former National Security Advisor, Stephen Hadley, recommend:

The President should establish a White House-led review process involving a small number of OMB and NSC staff to assess the allocation of resources and substantive progress toward meeting his national security priorities, and to recommend when course corrections may be needed. Second, the Administration must find a way to deliberately and periodically assess its strategy and how resources are aligned against it .... There are a number of ways to structure this assessment. One approach would be to bi-annually hold a Deputies-level and Principals-level dialogue on strategy efficacy to date.... [16]

Although the authors in this book recognize there is no silver-bullet solution to the strategy-resourcing misalignment the United States is currently facing, the president does have control of the National Security Council agenda. By creating a demand signal that the reconciliation of the ends and means of national strategy is a priority, there can at least be assurances of a periodic review of the process against established metrics of success. Implementing this recommendation will not solve the issues of the congressional committee structure, nor will it integrate the web of overlapping priorities among the DoD, DOS, and DHS, but it could begin to signal the administrations' resolve to improve the current process.

Recommendation 3: Shift Towards a Broader Conception of National Security

Twenty-first- century threats to US national security are not the same as the ones faced fifty or 100 years ago. In 1947 when Congress passed the National Security Act, national security was as an issue for the army and navy, and the geography of the problem set was overwhelmingly overseas. This is no longer true. In fact, cyber security, a key element of national security in the twenty-first century, has no geography at all. Thus, the current conception of neat divisions among departments and agencies, among federal and state, and between state and local governments no longer works. Nor does the division between government and private industry.

The COVID-19 pandemic that upended the country for more than year came at a cost to the American government in excess of $5 trillion for COVID relief.[17] Further, as of the time of this writing, the death toll from the pandemic has now exceeded that of total US casualties of World War II, the Korean War, and the Vietnam War combined.[18] Therefore, it is not an exaggeration to state that pandemics represent grave national security issues in multiple respects. That said, when actually faced with the pandemic, a risk recognized in the national security strategies of both the George W. Bush and Obama administrations, there were few tools in the American defense arsenal that could have been brought to bear to combat the effects of the disease on American citizens.[19]

The idea that the US government has over-resourced the Department of Defense while ignoring other security threats is not new. Many authors in this book have highlighted the disparities in funding among the agencies tasked with ensuring the security and prosperity of the country. Chapter 11, which details the resourcing conundrum associated with funding the Department of Homeland Security is an exemplar of the larger problem.

The current National Defense Strategy is built upon a premise of great power competition.[20] The United States is indeed facing strategic competition from near-peer adversaries that requires sober calculation of both the risks involved and the appropriate strategic responses required. That said, simply increasing military spending is insufficient.

In addition to military competition, in chapter 11, Mark Troutman outlines the real threats to critical infrastructure and intellectual property that are hallmarks of great power competition in the twenty-first century. To successfully compete in this new era requires that resourcing national security includes programs to combat climate change, disinformation, and election interference as well as the protection of intellectual property and cyber infrastructure.

Broadening the conception of national security in this way will require tradeoffs, and the most likely bill payer for this recommendation is the Department of Defense. That said, DoD spending currently accounts for nearly 50% of all US government discretionary spending. While the military threats facing the United States are real, they do not account for half of the national security threats the US currently faces. The concept of national security needs to be broadened, and spending must be rebalanced.

Recommendation 4: Facing the Hard Choices Necessary to Cut Unsustainable Deficit Spending

It doesn't matter whether your politics are liberal or conservative, there is a large math problem at the heart of the federal budget process, which was discussed in the section under "Things that Cannot Be Changed" in this chapter. The United States is consistently spending more than it takes in. This is causing a crowding out within the discretionary budget. The current contours of the debate—defense or everything else—blur the fact that the amount of discretionary spending—as a percent of the federal budget—decreases on an annual basis. Although given the current political climate, it is hard to see a way out of this conundrum. Until this cycle is broken, resourcing national security and, for that matter, everything else will continue to be problematic. The massive debt is an unchangeable fact, but the future trajectory can be altered if the political will is present.

As Mark Troutman points out in chapter 2, it is insufficient to cast this problem as a mere choice between "guns and butter." While historically low interest rates have allowed the United States breathing room, it has done nothing to correct the current unsustainable spending trajectory it finds itself on. Further, describing this situation as a relatively simply choice between "guns and butter" obscures the complexity of the problem.

CONCLUSION

So, where does that leave the reader who is earnestly searching for a means to rectify the fiscal imbalance and develop a more rational means to resource American national security priorities? As the editors, Mark

Troutman and I offer these thoughts. First, the system is inefficient and was designed to operate inefficiently. Despite this, the United States has made remarkable recoveries from initial defeats during times of crisis. From Pearl Harbor to the "Arsenal of Democracy" and again in the response to the 9/11 attacks, the rapid and comprehensive shifts in policy and strategy were undergirded by equally fantastic shifts in resourcing capacities and mandates. Second, although it is possible to make shifts in the system if one understands how the process works and has the patience to integrate strategy and resourcing from the outset, large shifts —absent an overwhelming crisis—which produce temporary consensus within the government are unlikely.

That said, crises seem to come along with alarming regularity. In chapter 10, John Ferrari talks about time as a critical factor in the resourcing calculus. He speaks of time both in terms of the amount of studying required to become an expert of the system and in terms of knowing when an idea or strategy has the greatest chance of actually being adopted and resourced. The first two decades of the twenty-first century were bookended by crises—the 9/11 attacks and the COVID-19 pandemic. Both required massive federal spending to ensure American national security. Both crises also represented an opportunity to reframe strategy and adjust priorities. The long-term effects of the pandemic remain as yet unknown as does the total cost in terms of dollars, lives, and human suffering. Perhaps the magnitude of the toll will provide an opening to rebalance national strategic resourcing.

NOTES

1. Reinhold Niebuhr, "The Serenity Prayer," *The Prayer Foundation*, (2010), https://www.prayerfoundation.org/dailyoffice/serenity_prayer_full_ version.htm.
2. There has been considerable controversy as to whether Reinhold Niebuhr is the actual author, or whether the prayer was adapted from another source. Saint Francis of Assisi is often credited. Further, Niebuhr's daughter, who believes that Niebuhr composed the prayer, cites it creation in the early 1940s. That said, multiple historians and biographers have credited Niebuhr with the prayer and have dated it back to the early 1930s through copies of sermons from that time. Regardless of its provenance, it is the sentiment of the prayer that remains relevant to this conclusion. For further information, see Brown, *Niebuhr and His Age*; and Niebuhr, *The Essential Reinhold Niebuhr*.
3. In *Federalist*, no. 51, James Madison wrote, "This policy of supplying, by opposite and rival interests, the defect of better motives, might be traced through the whole system of human affairs, private as well as public. We see it particularly displayed in all the subordinate distributions of power, where the constant aim is to divide and arrange the several offices in such a manner as that each may be a check on the other that the private interest of every individual may be a sentinel over the public rights." The full text is available from *The Avalon Project: Documents in Law, History and Diplomacy,* https://avalon.law.yale.edu/18th_century/fed10.asp
4. See Thomas McNaugher, chapter 6, "The Defense Budget Process."
5. All of the US National Security Strategies published are available online at the US National Security Strategy Archive at https://nssarchive.us.
6. "The U.S Constitution: Preamble," The United States Federal Courts, https://www.uscourts.gov/about-federal-courts/educational-resources/ about-educational-outreach/activity-resources/us.
7. Komer, *Bureaucracy Does Its Thing,*152.
8. "The National Security Act of 1947," Public Law 253, 80th Congress; Chapter 343, 1st Session; S. 758. July 26, 1947, https://global.oup.com/us/ companion.websites/9780195385168/resources/chapter10/nsa/nsa.pdf.
9. Thomas McNaugher, chapter 6, "The Defense Budget Process."
10. Demarest and Borghard, eds., *US National Security Reform*, 207; and Heidi Demarest, chapter 3, "The Role of Congress."

11. Kitrosser, "Congressional Oversight of National Security Activities,"107.

12. Geoffrey Odlum, chapter 5, "Determining and Resourcing Diplomatic Priorities."

13. Ibid.

14. Jason Galui, chapter 4, "A View from On High."

15. Geoffrey Odlum, chapter 5, Determining and Resourcing Diplomatic Priorities."

16. Hadley, Miller, and Carlin, *Connecting Strategy and Resources in National Security*, 10.

17. Peterson, "Here's Everything the Federal Government Has Done To Respond To The Coronavirus So Far."

18. Heather Hollingsworth and Tammy Webber, "U.S. Tops 500,000 Virus Deaths, Matching the Tolls of 3 Wars," *U.S. News and World Report,* February 21, 2021, https://www.usnews.com/news/health-news/articles/2021-02-22/vaccine-efforts-redoubled-as-us-death-toll-draws-near-500k.

19. In 2005, President George W. Bush signed a National Strategy for Pandemic Influenza; see https://georgewbush-whitehouse.archives.gov/homeland/pandemic-influenza-implementation.html. President Obama's 2015 National Security Strategy included a section titled "Increase Global Health Security," that dealt with the threat of and response to global pandemics. https://obamawhitehouse.archives.gov/sites/default/files/docs/2015_national_security_strategy_2.pdf.

20. "Summary of the 2018 National Defense Strategy of the United States of America," Office of the Secretary of Defense, 2, https://DoD.defense.gov/Portals/1/Documents/pubs/2018-National-Defense-Strategy-Summary.pdf.

Bibliography

Adams, Gordon, and Williams, Cindy *Buying National Security*: How America Plans and Pays for its Global Role and Safety at Home. New York: Routledge, 2010.

Allison, Graham T. *Essence of Decision.* New York: Little Brown, 1971.

Amadeo, Kimberly. "The U.S. Defense Department's Impact Is Bigger Than You Realize," *The Balance,* November 10, 2020. http://www.thebalance.com/department-of-defense-what-it-does-and-its-impact-3305982.

"America's Automobile Industry Is One of the Most Powerful Engines Driving the US Economy." Alliance of Automobile Manufacturers. October 4, 2019. https://autoalliance.org/economy/.

American Association for the Advancement of Science. "Historical Trends in Federal R&D." Accessed March 14, 2021. https://www.aaas.org/programs/r-d-budget-and-policy/historical-trends-federal-rd.

Arnold, R. Douglas. *The Logic of Congressional Action.* Philadelphia: University of Pennsylvania Press, 2011.

Art, Robert J., Vincent Davis, and Samuel P. Huntington, eds. *Reorganizing America's Defense: Leadership in War and Peace.* Newark, NJ: Pergamon- Brassey's, 1985.

Atkinson, Robert D., and Caleb Foote. "Federal Support for R&D Continues Its Ignominious Slide." Information Technology and Innovation Foundation. August 12, 2019. https://itif.org/publications/2019/08/12/federal-support-rd-continues-its-ignominious-slide#:~:text=The%20United%20States%20is%20continuing,research%20and%20development%20(R%26D).&text=At%20this%20pace%2C%20ITIF%20estimates,in%20R%26D%20investment%20by%202021.

Auerbach, Allen J., and William Gale, "The Effects of the COVID Pandemic on the Federal Budget Outlook." *Business Economics* 55 (October 2020): 202–212. https://doi.org/10.1057/s11369-020-00188-y.

Baumgartner and Jones, *The Politics of Information*: Problem Definition and the Course of Public Policy in American. Chicago: The University of Chicago Press, 2015.

Beardsley, Kyle. "Peacekeeping and the Contagion of Armed Conflict." *The Journal of Politics* 73, no. 4 (2011): 1051–64. https://www.journals. uchicago.edu/loi/jop.

———, and Kristian Skrede Gleditsch. "Peacekeeping as Conflict Containment." *International Studies Review* 17, no. 1 (2015): 67–89. https://academic.oup.com/isr.

Bedford, Bruce, Rand Beers, Chester Crocker, Karen Hanrahan, and David Miller. *State Department Reform Report.* Washington, DC: Atlantic Council, September 6, 2017. https://www.atlanticcouncil.org/in-depth-research-reports/report/state-department-reform-report/.

Beecroft, Robert M., and John Nalan. *Strengthening the Department of State.* Washington, DC: American Academy of Diplomacy, May 2019. https://www.academyofdiplomacy.org/wp-content/uploads/2021/01 /AADStrengtheningState.pdf.

Belasco, Amy. *The Cost of Iraq, Afghanistan, and Other Global War on Terror Operations Since 9/11.* December 8, 2014. Washington, DC: Congressional Research Service. https://sgp.fas.org/crs/natsec/RL3 3110.pdf.

Betts, Richard K. *Military Readiness: Concepts, Choices, Consequences.* Washington, DC: Brookings Institution, 1995.

Bhutada, Govind. "The U.S. Share of the Global Economy Over Time." Visual Capitalist, January 14, 2021. https://www.visualcapitalist.com/ u-s-share-of-global-economy-over-time/.

Blackwill, Robert D., and Jennifer M. Harris. *War by Other Means: Geoeconomics and Statecraft.* Cambridge, MA: The Belknap Press of Harvard University Press, 2017.

Board of Governors of the Federal Reserve System. "Credit and Liquidity Programs and the Balance Sheet." April 2, 2021. https://www. federalreserve.gov/monetarypolicy/bst_recenttrends.htm.

———. "Recent Balance Sheet Trends." https://www.federalreserve.gov/monetarypolicy/bst_recenttrends.htm.

Bonds, Timothy M, Michael Mazaar, James Dobbins, Michael Lostumbo, Michael Johnson, David A. Shlapak, Jeffrey Martini, et al., *America's Strategy - Resource Mismatch: Addressing the Gaps between U.S. National Strategy and Military Capacity.* Washington, DC: RAND, 2019. https://www.rand.org/content/dam/rand/pubs/research_reports/RR2100/RR2173z2/RAND_RR2173z2.pdf.

Bowie, Robert R., and Richard H. Immerman. *Waging Peace: How Eisenhower Shaped an Enduring Cold War Strategy.* New York: Oxford University Press, 1998.

British Broadcasting Corporation. "US Loses AAA Credit Rating after S&P Downgrade." August 6, 2011. https://www.bbc.com/news/world-us-canada-14428930.

Brown, Charles C. *Niebuhr and His Age: Reinhold Niebuhr's Prophetic Role and Legacy.* Harrisburg, PA: Trinity Press International, 2002.

Buchanan, Benjamin. *The Hacker and the State: Cyber Attacks and the New Normal of Geopolitics.* Cambridge, MA: Harvard University Press, 2020.

Bureau of Economic Analysis. "Gross Domestic Product, Fourth Quarter and Year 2020 (Second Estimate)." February 25, 2021. https://www.bea.gov/news/2021/gross-domestic-product-fourth-quarter-and-year-2020-second-estimate.

Bush, George W. *Securing the Homeland, Strengthening the Nation.* Washington, DC: Department of Homeland Security, 2003.

Cancian, Mark F. *Goldwater-Nichols 2.0.* Washington, DC: Center for Strategic and International Studies, March 4, 2016. https://www.csis.org/analysis/goldwater-nichols-20.

———, and Adam Saxton. *Industrial Mobilization: Assessing Surge Capabilities, Wartime Risk, and System Brittleness.* Washington, DC: Center for Strategic and International Studies, January 8, 2021. https://tinyurl.com/s9j24f6w.

Caro, Robert A. *The Power Broker: Robert Moses and the Fall of New York*. New York, NY: Vintage Books, 1975.

Center for Strategic and International Studies. "The Price of Peacekeeping" "Into Africa" podcast. April 15, 2021. https://www.csis.org/node/60583.

Chairman of the Joint Chiefs of Staff, Joint Capabilities Integration and Development System: CJCSI 3170.01H. Washington, DC: Department of Defense, 2012. https://www.acqnotes.com/Attachments/CJCSI%203170.01H%20Joint%20Capabilities%20Integration%20and%20Development%20System%2010%20January%202012.pdf.

Ciampoli, Paul. "Grid Exercise Evolves as Public Power Participation Increases." *American Public Power Association News*. November 14, 2019. https://www.publicpower.org/periodical/article/grid-exercise-evolves-public-power-participation-increases.

Clausewitz, Carl Von. *On War*. Edited by Michael Howard and Peter Paret. Princeton, NJ: Princeton University Press, 1976.

Cleveland, Charles, Benjamin M. Jensen, Arnel David, and Susan F. Bryant. *Military Strategy for the 21st Century: People, Connectivity, and Competition*. Amherst, NY: Cambria Press, 2018.

CNN Money. "Bush Seeks Up to 75B." October 3, 2001, https://money.cnn.com/2001/10/03/economy/economy_bush/

Cogan, John F. *The High Cost of Good Intentions: A History of U.S. Federal Entitlement Programs*. Stanford, California: Stanford University Press, 2019.

Coleman, Katharina P. "Extending UN Peacekeeping Financing Beyond UN Peacekeeping Operations? The Prospects and Challenges of Reform," *Global Governance* 23, no. 1 (2017): 101–120.

Congressional Budget Office. *The Budget and Economic Outlook: 2021 to 2031*. https://www.cbo.gov/system/files/2021-02/56970-Outlook.pdf.

———. "Discretionary Spending Options." https://www.cbo.gov/content/discretionary-spending-options#:~:text=Discretionary%20spending%E2%80%94the%20part%20of,35%20percent%20of%20federal%20outlays

———. "History." https://www.cbo.gov/about/history#:~:text=The%20 Budget%20and%20Accounting%20Act,the%20Budget%20(renamed%2 0the%20Office.

Cooper, Timothy, and Russell Rumbaugh. "Real Acquisition Reform." *Joint Forces Quarterly* 55, no. 4th Quarter (2009): 59–65. https://ndupress. ndu.edu/portals/68/Documents/jfq/jfq-55.pdf.

Cordesman, Anthony. *US Strategy, Sequestration, and the Growing Strategy-Reality Gap*. Washington, DC: Center for Strategic and International Studies, March 11, 2013. https://www.csis.org/analysis/ us-strategy-sequestration-and-growing-strategy-reality-gap.

Corera, Gordon. "North Korea Accused of Hacking Pfizer for Covid-19 Vaccine Data." *British Broadcasting Corporation*, February 16, 2021. https://www.bbc.com/news/technology-56084575.

Council of Economic Advisors. *Economic Report of the President*. Washington, DC: U.S. Government Publishing Office, February 2020. https://www.govinfo.gov/app/collection/erp/2020.

———. *Economic Report of the President*. Washington, DC: U.S. Government Publishing Office, February 2021. https://www.govinfo.gov/app/ collection/erp/2021.

Crews, Clyde Wayne. *Tip of the Costberg*. Washington, DC: Competitive Enterprise Institute, 2017. https://papers.ssrn.com/sol3/papers.cfm? abstract_id=2502883.

Crosson, Jesse M. et.al. "Partisan Competition and the Decline in Legislative Capacity among Congressional Offices." *Legislative Studies Quarterly*. July 29, 2020. https://doi.org/10.1111/lsq.12301.

Curry, Timothy, Lynn Shibut, and James A. Marino. "Cost of the Savings and Loan Crisis: Summary and Implications" *FDIC Banking Review* 13 no. 2 (January 2000): 26–35. http://www.workingre.com/wp-content/ uploads/2013/08/cost-of-SL.pdf.

Cybersecurity and Infrastructure Security Agency. "Cyber-Attack Against Ukrainian Critical Infrastructure." Department of Homeland Security. February 25, 2016. https://us-cert.cisa.gov/ics/alerts/IR-ALERT-H-1 6-056-01.

Daggett, Stephen. *Costs of Major U.S. Wars*. June 29, 2010. Washington, DC: Congressional Research Service. https://sgp.fas.org/crs/natsec/RS22926.pdf.

Davis, Thomas. *Framing the Problem of PPBS*. Washington, DC: Business Executives for National Security, 2000. https://www.bens.org/Publications.

Demarest, Heidi. *US Defense Budget Outcomes*: Volatility and Predictability in Army Weapons Funding. London: Palgrave Macmillan, 2017.

Demarest, Heidi B., and Erica D. Borghard, eds. *US National Security Reform: Reassessing the National Security Act of 1947*. New York: Routledge, 2019.

Deni, John R. *The Future of American Landpower: Does Forward Presence Still Matter? The Case of the Army in the Pacific*. Carlisle, PA: Strategic Studies Institute, 2014. https://press.armywarcollege.edu/cgi/viewcontent.cgi?article=1490&context=monographs.

———. *Rotational Deployments vs. Forward Stationing: How Can the Army Achieve Assurance and Deterrence Efficiently and Effectively*. Carlisle, PA: USAWC Strategic Studies Institute, 2017. https://press.armywarcollege.edu/monographs/408/.

Diebel, Terry L. *Foreign Affairs Strategy*. New York, NY: Cambridge University Press, 2007.

"DOD's Role in the Competition with China." Testimony Before the House Armed Services Committee, 116th Cong. (Testimony of Michele A. Flournoy). https://armedservices.house.gov/_cache/files/4/4/44fbef3d-138c-4a0a-b3a9-2f05c898578f/0E4943A5BFAEDA465D485A166FABCF5F.20200115-hasc-michele-flournoy-statement-vfinal.pdf

Dorman, Shawn. *Inside a U.S. Embassy: Diplomacy at Work*. Washington, D.C.: Foreign Service Books, 2011.

Doyle, Michael W., and Nick Sambanis. "International Peacebuilding: A Theoretical and Quantitative Analysis." *American Political Science Review* 94, no. 4 (2000): 779–801.

"Economics Terms A to Z." *The Economist*. Accessed April 7, 2021. https://www.economist.com/economics-a-to-z.

Eisenhower, Dwight D. "Farewell Address." Dwight D. Eisenhower Presidential Library, January 17, 1961, audio file, 5:15. https://www.eisenhowerlibrary.gov/sites/default/files/all-about-ike/speeches/wav-files/farewell-address.mp3.

Enthoven, Alan C., and K. Wayne Smith. *How Much Is Enough: Shaping the Defense Program 1961-1969*. New York, NY: Harper and Row, 1971.

Federal Emergency Management Agency. *National Preparedness Goal* (First Edition, September 2011). Department of Homeland Security, 2011. https://www.fema.gov/pdf/prepared/npg.pdf.

Flournoy, Michèle, and Janine Davidson. "Obama's New Global Posture: The Logic of U.S. Foreign Deployments." *Foreign Affairs* 91, no. 4 (2012): 54–63. https://www.jstor.org/stable/23218039?seq=1.

Foreign Relations Authorization Act, Fiscal Year 2003, Public Law 107–228 (September 30, 2003). https://www.congress.gov/107/plaws/publ2 28/PLAW-107publ228.pdf

Fortna, Virginia Page. *Does Peacekeeping Work? Shaping Belligerents' Choices After Civil War*. Princeton: Princeton University Press, 2008.

Fox, J. Ronald, David Grayson Allen, Thomas Charles Lassman, Walton S. Moody, and Philip Shiman. *Defense Acquisition Reform 1960–2009: An Elusive Goal*. Washington, DC: Center for Military History, 2015.

Friedman, Joel, Sharon Parrott, and Aviva Aaron-Dine. "Five Things to Look for in the New Trump Budget," Washington, DC: Center on Budget and Policy Priorities, March 8, 2019. https://www.cbpp.org/research/federal-budget/five-things-to-look-for-in-the-2020-trump-budget.

Friedman, Milton, and Anna Jacobson Schwartz. *A Monetary History of the United States: 1867-1960*. Princeton, NJ: Princeton University Press, 1993.

Fullweiler, Scott T. "The Debt Ratio and Sustainable Economic Policy." *World Economic Review* 41, no. 7 (December 2007): 1003–1042.

Furman, Jason, and Lawrence H. Summers. "A Reconsideration of Fiscal Policy in the Era of Low Interest Rates." Discussion Draft. https://

www.piie.com/system/files/documents/furman-summers2020-12-01
paper.pdf.

Furnas, Alexander C. and Timothy N. LaPira. "Congressional Brain
Drain: Legislative Capacity in the 21st Century." *New America*
September 2020. https://d1y8sb8igg2f8e.cloudfront.net/documents/
Congressional_Brain_Drain.pdf.

Gain, Nathan. *U.S. Navy Receives 1st Virginia-Class Block IV Nuclear-
Powered Attack Submarine From GDEB.* April 18, 2020. https://www.
navalnews.com/naval-news/2020/04/u-s-navy-receives-1st-virginia-
class-block-iv-nuclear-powered-attack-submarine-from-gdeb/.

Gates, Robert M. *Duty: Memoirs of A Secretary at War.* New York, NY:
Vintage Books, 2015.

———. "Secretary of Defense Speech to the Corps of Cadets." Stars and
Stripes, February 25, 2011. https://www.stripes.com/news/text-of-
secretary-of-defense-robert-gates-feb-25-2011-speech-at-west-point-
1.136145

"GDP per Capita (Current US$) - United States." World Bank. Accessed
March 10, 2021. https://data.worldbank.org/indicator/NY.GDP.PCAP.
CD?locations=US.

George, Roger Z., and Harvey Rishikof. *The National Security Enterprise:
Navigating the Labyrinth.* Washington, DC: Georgetown University
Press, 2017.

Gladwell, Malcolm. *Outliers: The Story of Success.* New York, NY: Back
Bay Books, 2011.

Gordon, John Steele. *An Empire of Wealth: The Epic History of American
Economic Power.* New York, NY: Harper Perennial, 2005.

Goss, Carol F. "Military Committee Membership and Defense-Related
Benefits in the House of Representatives." *Western Political Quarterly*
25, no. 2 (1972): 215–233.

Goss, Thomas. "Who's In Charge? New Challenges in Homeland Defense
and Homeland Security." *Homeland Security Affairs*, 2006, 1–11.

Gould, Joseph. "Wide Reaching Hack Has Defense Firms on Their Toes."
C4ISRNet, December 23, 2020. https://www.c4isrnet.com/2020/12/23/

wide-reaching-hack-has-defense-firms-on-their-toes/. Gray, Colin S. *The Strategy Bridge: Theory for Practice*. Oxford, UK: Oxford University Press, 2010.

Greenberg, Allen. *Confessions of A Government Man: How to Survive in Any Bureaucracy*. Indianapolis: Dog Ear Publishing, 2020.

Greenspan, Alan, and Adrian Wooldridge. *Capitalism in America*. New York: Penguin, 2019.

Greenwald, Michael. *The Future of the United States Dollar*. Washington, DC: Atlantic Council, December 8, 2020. https://www.atlanticcouncil. org/in-depth-research-reports/the-future-of-the-united-states-dollar/ . ruber, Jonathan, and Simon Johnson. *Jump-Starting America: How Breakthrough Science Can Revive Economic Growth and the American Dream*. New York: Public Affairs, 2019.

Hadley, Stephen J., James N. Miller, and Mara E. Carlin, *Connecting Strategy and Resources in National Security: Recommendations for Trump Administration*. *The Atlantic Council*, February 28, 2019. https://www.atlanticcouncil.org/content-series/strategy-consortium/connecting-strategy-and-resources-in-national-security-10-recommendations-for-the-trump-administration/.

Heller, Charles E., and William A. Stofft, eds. *America's First Battles: 1776–1965*. Lawrence: University of Kansas Press, 1986.

Henderson, David R. "Economic Growth." In *The Concise Encyclopedia of Economics*. Indianapolis, IN: Liberty Fund, 2008.

Hendrix, Jerry, and Benjamin Armstrong. *The Presence Problem: Naval Presence and National Strategy*. Washington, DC: Center for a New American Security, January 2016. https://s3.amazonaws.com/files. cnas.org/documents/The_Presence_Problem_FINAL.pdf?mtime=20 160906082551.

Heniff, Bill Jr. *Formulation and Content of the Budget Resolution*. May 1, 2007. Washington, DC: Congressional Research Service. https://www. senate.gov/CRSpubs/0913b27b-60a2-4f12-a4ca-80fbfc65b049.pdf.

Herman, Arthur. *Freedom's Forge: How American Business Produced Victory in World War II*. New York: Random House Trade Paperbacks, 2013.

"Historical National Accounts for the United States." Groningen Growth and Development Centre, 2009. https://www.rug.nl/ggdc/historicaldevelopment/na/.

Hoffman, Francis G. "Grand Strategy: The Fundamental Considerations." *Orbis* 58, no. 4 (2014): 472–85. https://doi.org/10.1016/j.orbis.2014.08.002.

How the Army Runs: A Senior Leader Reference Handbook, 2020-21. Carlisle Barracks, PA: U.S. Army War College, 2020. https://publications.armywarcollege.edu/pubs/3736.pdf.

Hsu, Jeremy. "Pentagon Warns Silicon Valley About Aiding Chinese Military." *IEEE Spectrum.* March 28, 2019. https://spectrum.ieee.org/pentagon-warns-silicon-valley-about-aiding-chinese-military#toggle-gdpr.

Ingram, George. "What Every American Should Know About US Foreign Aid." Brookings Institution, October 1, 2020. https://www.brookings.edu/policy2020/votervital/what-every-american-should-know-about-us-foreign-aid/.

Interagency Task Force in Fulfillment of Executive Order 13806. *Assessing and Strengthening the Manufacturing and Defense Industrial Base and Supply Chain Resiliency of the United States.* Washington, DC. September 2018. https://media.defense.gov/2018/Oct/05/2002048904/-1/-1/1/ASSESSING-AND-STRENGTHENING-THE-MANUFACTURING-AND-DEFENSE-INDUSTRIAL-BASE-AND-SUPPLY-CHAIN-RESILIENCY.PDF.

Johnson, Bridget. "Three-Year CFATS Reauthorization Signed Just Before DHS Program Was Set to Expire." *Homeland Security Today,* July 22, 2020. https://www.hstoday.us/subject-matter-areas/infrastructure-security/three-year-cfats-reauthorization-signed-just-before-dhs-program-was-set-to-expire/.

Junor, Laura J. *Managing Military Readiness. Strategic Perspectives,* no. 23. Washington, DC: NDU Press, 2017. https://ndupress.ndu.edu/Portals/68/Documents/stratperspective/inss/Strategic-Perspectives-23.pdf.

Kaufman, Herbert. *The Administrative Behavior of Federal Bureau Chiefs.* Washington, DC: Brookings Press, 1981.

Kennedy, Paul. *The Rise and Fall of the Great Powers: Economic Change and Military Conflict from 1500–2000*. New York, NY: Vintage Books, 1989.

Kessler, Glenn. "Did Wall Street Get a 'Trillion-Dollar Bailout' during the Financial Crisis?" *Washington Post*. March 8, 2019. https://www. washingtonpost.com/politics/2019/03/18/did-wall-street-get-trillion-dollar-bailout-during-financial-crisis/.

Kiel, Paul, and Daniel Nguyen. "Bailout Tracker." *ProPublica*. February 16, 2021. https://projects.propublica.org/bailout/list.

Kiewiet, D. Roderick, and Mathew D. McCubbins. "Congressional Appropriations and the Electoral Connection." *Journal of Politics* 47, no. 1 (1985): 59–72.

Kim, Young Eun, and Norman V Loayza. *Productivity Growth: Patterns and Determinants across the World*. World Bank, May 29, 2019. https:// elibrary.worldbank.org/doi/abs/10.1596/1813-9450-8852.

King, Angus, and Michael Gallagher. "Commission Report." Washington, DC: Cyberspace Solarium Commission, March 2020. https://www. solarium.gov/report.

Kitrosser, Heidi. "Congressional Oversight of National Security Activities: Improving Information Funnels." *Cardozo Law Review* 29, no. 3 (2008): 1049–1090.

Komer, Robert W. *Bureaucracy Does Its Thing: Institutional Constraints on U.S.-GVN Performance in Vietnam*. Santa Monica, CA: RAND Corporation, 1972. https://www.rand.org/pubs/reports/R967.html.

Kravitz, Walter. "The Advent of the Modern Congress: The Legislative Reorganization Act of 1970," *Legislative Studies Quarterly* 15, no. 3 (1990): 375-399.

Lane, Carl. "The Elimination of the National Debt in 1835 and the Meaning of Jacksonian Democracy," *Essays in Economic and Business History* 25 (2007): 67–78.

Larson, Eric V. *Force Planning Scenarios, 1945–2016: Their Origins and Use in Defense Strategic Planning*. Santa Monica, CA: RAND, 2019. https:// www.rand.org/pubs/research_reports/RR2173z1.html.

———, Kristin Leuschner, and David T. Orletsky. *Defense Planning in a Decade of Change: Lessons from the Base Force, Bottom-up Review, and Quadrennial Defense Review*. Santa Monica, CA: RAND, 2001. https://www.rand.org/pubs/monograph_reports/MR1387.html.

Leiner, Barry M, Vinton G. Cerf, David D. Clark, Robert E. Kahn, Leonard Kleinrock, Daniel C. Lynch, Jon Postel, Larry G. Roberts, and Stephen Wolff. "Brief History of the Internet." Internet Society, August 14, 2020. https://www.internetsociety.org/internet/history-internet/brief-history-internet/.

Leslie, Keith J., and Max P Michaels. "The Real Power of Real Options." *McKinsey Quarterly* 3 (June 1, 2000). https://www.mckinsey.com/business-functions/strategy-and-corporate-finance/our-insights/the-real-power-of-real-options.

Lewis, James Andrew. "The Economic Impact of Cybercrime." Washington, DC: Center for Strategic and International Studies, February 21, 2018. https://www.csis.org/analysis/economic-impact-cybercrime.

Lewis, T. G. *Critical Infrastructure Protection in Homeland Security: Defending a Networked Nation*. Hoboken, NJ: John Wiley & Sons Inc., 2020.

Liebman, Jeffrey B., and Neale Mahoney. "Do Expiring Budgets Lead to Wasteful Year-End Spending? Evidence from Federal Procurement." National Bureau of Economic Research Working Paper 19481. September 2013. Revised January 2018.

Locher, James R. *Project on National Security Reform: Forging a New Shield*. Washington, DC: Project on National Security Reform, November 2008. https://apps.dtic.mil/dtic/tr/fulltext/u2/a491826.pdf.

Loiselle, Marie-Eve. "The penholder system and the rule of law in the Security Council decision-making: Setback or improvement?" *Leiden Journal of International Law* 33 (2020): 139–156.

Lostumbo, Michael J. et al., *Overseas Basing of U.S. Military Forces: An Assessment of Relative Costs and Strategic Benefits*. Santa Monica, CA: RAND Corporation, 2013. https://www.rand.org/pubs/research_reports/RR201.html.

Luttwak, Edward. *The Grand Strategy of the Roman Empire: From the First Century CE to the Third.* Baltimore: Johns Hopkins University Press, 2016.

Lykke, Arthur F. "Defining Military Strategy." *Military Review* 69, no. 5 (May 1989): 2–8.

Lynn, Larry, and Richard I. Smith, "Can the Secretary of Defense Make a Difference?" *International Security* 7, no. 1 (Summer 1982): 45–69.

MacColl, Jamie, and Sneha Dawda. "US Water Plant Suffers Cyber Attack Through the Front Door." *RUSI Journal.* February 10, 2021. https://rusi.org/commentary/us-water-plant-suffers-cyber-attack-through-front-door.

Maddison, Angus. *The World Economy: Historical Statistics.* Paris: OECD Development Centre, 2003.

Mankiw, N. Gregory. *Macroeconomics.* 10th ed. New York: Worth Publishers, 2018.

Manyika, James, Susan Lund, Jacques Bughin, Jonathan Woetzel, Kalin Stamenov, and Dhruv Dhingra. "Digital Globalization: The New Era of Global Flows." *McKinsey Global Institute*, February 24, 2016. https://www.mckinsey.com/business-functions/mckinsey-digital/our-insights/digital-globalization-the-new-era-of-global-flows.

Mayer, Kenneth B. *The Political Economy of Defense Contracting.* New Haven, CT: Yale University Press, 1991.

Mayhew, David R. *Congress: The Electoral Connection.* New Haven: Yale University Press, 2006.

Mazaar, Michael J., Katharina Ley Best, Burgess Laird, Michael E. Linick, and Dan Madden. *The U.S. Department of Defense's Planning Process: Components and Challenges.* Santa Monica, CA: RAND, 2019. https://www.rand.org/pubs/research_reports/RR2173z2.html.

McCubbins, Matthew D., and Thomas Schwartz, "Congressional Oversight Overlooked: Police Patrols versus Fire Alarms," *American Journal of Political Science* 28, no. 1 (1984): 165–179.

Meiser, Jeffrey W. "Are Our Strategic Models Flawed? Ends + Ways + Means = Bad Strategy." *Parameters* 46, no. 4 (2016): 81–92.

Mian, Atif, Ludwig Straub, and Amir Sufi, "A Goldilocks Theory of Fiscal Policy." National Bureau of Economic Research Working Paper. https://scholar.harvard.edu/files/straub/files/goldilocks.pdf.

Moretti, Enrico, Claudia Steinwender, and John Van Reenen. "The Intellectual Spoils of War? Defense R&D, Productivity and International Spillovers." Working Paper. National Bureau of Economic Research. November 2019 (Revised August 2020). https://www.nber.org/papers/w26483.

Morgan, Matthew J. *The Impact of 9/11 on Politics and War.* New York: Palgrave MacMillan, 2009.

National Constitution Center. "Happy Birthday to the Department of State." National Constitution Center, July 27, 2020. https://constitutioncenter.org/blog/happy-237th-birthday-to-the-department-of-state.

National Defense Authorization Act for Fiscal Year 2020, Public Law 116-92 (December 20, 2019). https://congress.gov/116/plaws/publ92/PLAW-116publ92.pdf.

Niebuhr, Reinhold. *The Essential Reinhold Niebuhr: Selected Essays and Addresses.* Edited by Robert MacAfee Brown. Yale University Press, 1987.

Office of Management and Budget. "Budget of the U.S. Government, Fiscal Year 2020." Washington, DC: US Government Publishing Office, March 18, 2019. https://www.govinfo.gov/features/budget-fy2020.

Office of the Historian. "The Department of State." Foreign Service Institute, 2020. https://history.state.gov/departmenthistory.

Office of the Inspector General. "Inspection of the Bureau of Near East Affairs." Department of State, 2017. https://www.stateoig.gov/system/files/isp-i-17-22.pdf

Office of the President. *National Security Strategy of the United States (2017).* Washington, DC: December 2017. https://trumpwhitehouse.archives.gov/wp-content/uploads/2017/12/NSS-Final-12-18-2017-0905.pdf.

Office of the Under Secretary of Defense (Comptroller). *National Defense Budget Estimates for FY 2020 (Green Book).* https://comptroller.defense.gov/Portals/45/Documents/defbudget/FY2020/FY20_Green_Book.pdf.

———. *National Defense Budget Estimates for FY 2021 (Green Book).* https://comptroller.defense.gov/Portals/45/Documents/defbudget/FY2021/FY21_Green_Book.pdf.

Paladino, Alex. "Top 100 Global Tech Leaders." *Thomson Reuters,* January 17, 2018. https://www.thomsonreuters.com/en/products-services/technology/top-100.html.

Peterson, Michael. "Here's Everything the Federal Government Has Done to Respond to the Coronavirus So Far." Peter G. Peterson Foundation (blog), March 10, 2021. https://www.pgpf.org/blog/2021/03/heres-everything-congress-has-done-to-respond-to-the-coronavirus-so-far.

———. "U.S. Defense Spending Compared to Other Countries." Peter G. Peterson Foundation, May 13, 2020. https://www.pgpf.org/chart-archive/0053_defense-comparison.

Pramanik, Abhik K. "The Case for Keeping USAID and the Department of State Separate." Blog. U.S. Global Leadership Coalition, April 10, 2017. https://www.usglc.org/blog/the-case-for-keeping-usaid-and-the-state-department-separate/.

———. "Reforming Diplomacy: Standing on the Shoulders of Giants." United States Global Leadership Coalition (USGLC) report. June 22, 2017. https://www.usglc.org/blog/reforming-diplomacy-standing-on-the-shoulders-of-giants/.

Rapp, Nicolas. "Visualizing the Global Fortune 500 (2020)." Mapping the Global 500. *Forbes,* 2020. https://interactives.fortune.com/global_500_2020/dashboard/index.html.

Readiness Reports, 10 U.S. Code § 482, Legal Information Institute. https://www.law.cornell.edu/uscode/text/10/482.

Rosenzweig, Paul. "Streamlining Congressional Oversight of DHS." *Lawfare Blog,* October 31, 2019. https://www.lawfareblog.com/streamlining-congressional-oversight-dhs.

Rostow, W. W. *The Stages of Economic Growth: A Non-Communist Manifesto*. Eastford, CT: Martino Fine Books, 2017.

Ruggieri, Andrea, Han Dorussen, and Theodora-Ismene Gizelis. "Winning the Peace Locally: UN Peacekeeping and Local Conflict." *International Organization* 71, no. 1 (2018): 163–85.

Rumbaugh, Russell. *What We Bought*. Washington, DC: Henry L. Stimson Center, October 2011. https://www.stimson.org/wp-content/files/file-attachments/Contentv2_1.pdf.

Sandler, Todd. "International Peacekeeping Operations." *Journal of Conflict Resolution* 61, no. 9, (2017): 1875–1897. https://doi.org/10.1177/0022002717708601

Sapolsky, Harvey M., Eugene Gholz, Caitlin Talmadge. *US Defense Politics: The Origins of Security Policy*. New York: Routledge, 2017.

Saturno, James V. *Biennial Budgeting: Issues, Options, and Congressional Actions*. Washington, DC: Congressional Research Service. January 10, 2017. https://sgp.fas.org/crs/misc/R44732.pdf.

———. "Introduction to the Federal Budget Process." Washington, DC: Congressional Research Service. R46240. February 26, 2020. https://crsreports.congress.gov/product/pdf/R/R46240.

Schick, Alan. *The Federal Budget: Policy, Politics, Process*. 2nd ed. Washington, DC: Brookings, 2000.

Schwab, Klaus, and Zahidi Zahidi. "Global Competitiveness Report 2020." World Economic Forum, December 16, 2020. https://www.weforum.org/reports/the-global-competitiveness-report-2020.

Schwarzkopf, H. Norman, and Peter Petre. *General H. Norman Schwarzkopf: The Autobiography: It Doesn't Take a Hero*. London: Bantam, 1992.

Sharland, Lisa. "How Peacekeeping Policy Gets Made: Navigating Intergovernmental Processes at the UN," New York: International Peace Institute, May 2018. https://www.ipinst.org/wp-content/uploads/2018/05/1805_How-Peacekeeping-Policy-Gets-Made.pdf.

Shlapak, David A., and Michael Johnson. *Reinforcing Deterrence on NATO's Eastern Flank: Wargaming the Defense of the Baltics*. Santa Monica,

CA: RAND, 2016. https://www.rand.org/pubs/research_reports/RR1
253.html.

Sims, Robert B. *The Pentagon Reporters.* Honolu, HI: University Press of
the Pacific, 2005.

Smith, Adam. "The United States of America" in *Providing Peacekeep-
ers: The Politics, Challenges, and Future of United Nations Peacekeeping
Contributions* edited by Alex J. Bellamy and Paul D. Williams, 71-93,
Oxford: OUP Oxford, 2014.

Smith, Adam C., and Arthur Boutellis, "Rethinking Force Generation:
Filling the Capability Gaps in UN Peacekeeping." *Providing for
Peacekeeping* 2 (May 2013), https://www.ipinst.org/wp-content/
uploads/publications/ipi_rpt_rethinking_force_gen.pdf.

Smith, Perry M., and Daniel M. Gerstein. *Assignment: Pentagon: How to
Excel in a Bureaucracy.* Lincoln, NE: Potomac Books, 2020.

Smits, J.P., P.J. Woltjer, and D. Ma. "A Dataset on Comparative Historical
National Accounts, Ca. 1870-1950: A Time-Series Perspective."
Groningen Growth and Development Centre, 2009. https://www.rug.
nl/ggdc/historicaldevelopment/na/.

Social Security Administration. "Monthly Statistical Snapshot, September
2021." https://www.ssa.gov/policy/docs/quickfacts/stat_snapshot/.

Stamborski, Alan J. "A Look at the Fed's Dual Mandate." Federal Reserve
Bank of St. Louis, October 30, 2018. https://www.stlouisfed.org/open-
vault/2018/august/federal-reserve-dual-mandate.

Steil, Benn. *The Battle of Bretton Woods John Maynard Keynes, Harry Dex-
ter, and the Making of a New World Order.* Princeton, New Jersey:
Princeton Univ. Press, 2014.

Summers, Lawrence H. "The Biden stimulus is admirably ambitious.
But it brings some big risks, too." *Washington Post,* February 4,
2021. https://www.washingtonpost.com/opinions/2021/02/04/larry-
summers-biden-covid-stimulus/. https://www.washingtonpost.com/
opinions/2021/02/04/larry-summers-biden-covid-stimulus/.

Sun Tzu, *The Art of War.* Translated by Michael Nylan. New York: W.W.
Norton and Company, 2020.

Swagel, Phillip. "The 2021 Long-Term Budget Outlook." Congressional Budget Office, March 4, 2021. https://www.cbo.gov/publication/56977.

"10 Artificial Intelligence Companies Leading the Smart Revolution." *Analytics Insight News,* August 31, 2019. https://www.analyticsinsight. net/10-artificial-intelligence-companies-leading-smart-revolution/.

Tellis, Ashley J. *Measuring National Power in the Postindustrial Age.* Santa Monica, CA: RAND/Arroyo Center, 2000. https://www.rand. org/pubs/monograph_reports/MR1110.html.

"The Overseas Activities of Chinese Banks Shift Up a Gear." *The Economist,* October 29, 2020. https://www.economist.com/finance- and-economics/2020/10/28/the-overseas-activities-of-chinese-banks- shift-up-a-gear.

Thomas, Clayton. "Afghanistan: Background and Policy." Washington, DC: Congressional Research Service. November 10, 2020. https://fas. org/sgp/crs/row/R45122.pdf.

Under Secretary of Defense (Comptroller), Department of Defense Financial Management Regulation, DoD 7000.14 § (2019). https:// comptroller.defense.gov/Portals/45/documents/fmr/Combined_ Volume1-16.pdf.

Undersecretary of Defense (Personnel and Readiness). "Deployment- to-Dwell, Mobilization-to-Dwell Revision." Accessed April 5, 2021. https://mccareer.files.wordpress.com/2016/11/deployment-to-dwell- instruction.pdf.

Under Secretary of Defense for Policy. *Management of U.S. Global Defense Posture, DoDI 3000.12 (with change 1).* Department of Defense (Washington, DC: May 8, 2017). https://www.esd.whs.mil/Portals/54 /Documents/DD/issuances/dodi/300012p.pdf.

UNESCO Institute for Statistics. "How Much Does Your Country Invest in R&D?" Accessed March 14, 2021. http://uis.unesco.org/apps/ visualisations/research-and-development-spending/.

"United States' Arms Exports Totaled $175.08 Billion in 2020, up 2.8% Over 2019." *Defense World,* December 5, 2020. https://www.defenseworld. net/news/28470/United_States____Arms_Exports_Totaled__175_08 _billion_in_2020__up_2_8__Over_2019#.YE0mwTqSlPZ.

U.S. Department of Defense. *Department of Defense Dictionary of Military and Associated Terms.* Washington, DC: 2013. https://code7700.com/pdfs/jp1_02.pdf.

———. "DOS and DOD Officials Brief Reporters on Fiscal 2020 Arms Transfer Figures." *Defense Newsroom.* December 4, 2020. https://www.defense.gov/Newsroom/Transcripts/Transcript/Article/2437110/dos-and-dod-officials-brief-reporters-on-fiscal-2020-arms-transfer-figures/.

———. "Statement on Strategic Choices and Management Review," July 31, 2013. https://archive.defense.gov/Speeches/Speech.aspx?SpeechID=1798.

———. *Summary of the 2018 National Defense Strategy of the United States of America: Sharpening the American Military's Competitive Edge.* 2018. https://dod.defense.gov/Portals/1/Documents/pubs/2018-National-Defense-Strategy-Summary.pdf.

U.S. Department of Defense Inspector General. *Procurement Quantities of the AH-64E Apache New Build and Remanufacture Helicopter Programs.* DODIG 2018-130 (Washington DC, 2018). https://media.defense.gov/2018/Jun/27/2001936583/-1/-1/1/DODIG-2018-130.PDF.

U.S. Department of Homeland Security. "Creation of the Department of Homeland Security." September 24, 2015. https://www.dhs.gov/creation-department-homeland-security.

———. "DHS Mission." July 3, 2019. https://www.dhs.gov/mission.

———. "Guiding Principles." July 3, 2019. https://www.dhs.gov/guiding-principles.

———. "The DHS SAFETY Act." Accessed March 13, 2021. https://www.safetyact.gov/.

U.S. Department of State. "Duties of the Secretary of State." 2020. https://www.state.gov/duties-of-the-secretary-of-state/.

———. *Foreign Affairs Manual and Handbook.* 2020. https://fam.state.gov/Default.aspx.

———. "Global Talent Management Fact Sheet." June 30, 2020. http://www.afsa.org/sites/default/files/0620_state_dept_hr_factsheet.pdf.

———. "Joint Strategic Plan: FY 2018-2022." February 2018. https://www.
state.gov/wp-content/uploads/2018/12/Joint-Strategic-Plan-FY-2018
-2022.pdf.

U.S. Government Accountability Office. *Biennial Budgeting: Three States'
Experiences.* October 2000. https://www.gao.gov/assets/gao-01-132.
pdf.

———. *Defense Acquisitions Annual Assessment: Drive to Deliver Capabilities
Faster Increases Importance of Program Knowledge and Consistent
Data for Oversight: Defense Acquisitions Annual Assessment 2020.*
GAO-20-439 Washington, DC: June 3, 2020). https://www.gao.gov/
products/gao-20-439.

———. *DHS Management: Enhanced Oversight Could Better Ensure Programs
Receiving Fees and Other Collections Use Funds Efficiently.* GAO-16-443
(Washington, DC: July 21, 2016). https://www.gao.gov/products/gao-
16-443.

———. *Homeland Security Grant Program: Additional Actions Could Further
Enhance FEMA's Risk-Based Grant Assessment Model,* September 6,
2018. https://www.gao.gov/products/gao-18-354.

———. *Homeland Security: DHS Risk-Based Grant Methodology Is
Reasonable, But Current Version's Measure of Vulnerability is Limited,*
June 27, 2008. https://www.gao.gov/products/gao-08-852.

U.S. Library of Congress. Congressional Research Service. *Acquisition
Reform in the FY2016-FY2018 National Defense Authorization Acts
(NDAAs),* by Heidi M. Peters. R45068. January 19, 2018. https://
crsreports.congress.gov/product/pdf/R/R45068.

———. *Base Closure and BRAC: Background and Issues for Congress,* by
Christopher T. Mann. R45705. April 25, 2019. https://crsreports.
congress.gov/product/pdf/R/R45705

———. *Continuing Resolutions: Overview of Components and Practices,* by
Kate P. McClanahan, James V. Saturno, and Megan S. Lynch. April
19, 2019. https://crsreports.congress.gov/product/pdf/R/R42647

———. *Defense Primer: Future Years Defense Program (FYDP),* by Heidi M.
Peters and Brendon W McGarry. IF10831. December 14, 2020. https://
crsreports.congress.gov/product/pdf/IF/IF10831.

———. *The Defense Production Act of 1950: History, Authorities, and Considerations for Congress*, by Michael H. Cecire and Heidi M. Peters. R43767. March 2, 2020. https://crsreports.congress.gov/product/pdf/R/R43767.

———. *Department of Defense Use of Other Transaction Authority: Background, Analysis, and Issues for Congress*, by Heidi M. Peters. February 22, 2019. https://crsreports.congress.gov/product/pdf/R/R45521.

———. *Department of State, Foreign Operations, and Related Programs: FY2020 Budget and Appropriations* by Cory R. Gill, Marian L. Lawson, and Emily M. Morgenstern. April 9, 2020. https://crsreports.congress.gov/product/pdf/R/R45763.

———. *Economics and National Security: Issues and Implications for U.S. Policy*, by Richard K. Nanto. R41589. January 4, 2011. https://fas.org/sgp/crs/natsec/R41589.pdf.

———. *Foreign Assistance: An Introduction to U.S. Programs and Policy*, by Marian Lawson and Emily Morgenstern. April 30, 2020. https://crsreports.congress.gov/product/pdf/R/R40213.

———. *Foreign Assistance Act of 1961: Authorizations and Corresponding Appropriations*, by Diane E Rennack and Susan G. Chesser. R40089. April 18, 2012. https://crsreports.congress.gov/product/pdf/R/R40089.

———. *FY 2021 Defense Budget Request: An Overview*, by Brendan W. McGarry. IN11224. February 20, 2020. https://crsreports.congress.gov/product/pdf/IN/IN11224

———. *Introduction to the Federal Budget Process*, by James V. Saturno. R46240. February 26, 2020. https://crsreports.congress.gov/product/pdf/R/R46240.

———. *Navy Virginia (SSN-774) Class Attack Submarine Procurement: Background and Issues for Congress*, by Ronald O'Rourke. RL32418. February 23, 2021. https://crsreports.congress.gov/product/pdf/RL/RL32418.

———. *Trends in the Timing and Size of DHS Appropriations: In Brief*, by William L. Painter. R44604. December 6, 2019. https://crsreports.congress.gov/product/pdf/R/R44604.

———. *U.S. Foreign Assistance: Budget Development and Execution*, by Nick Brown. April 23, 2020. https://crsreports.congress.gov/product/pdf/IF/IF11515.

———. *U.S. Funding to the United Nations System: Overview and Selected Policy Issues*, by Luisa Blanchfield. R45206. April 25, 2018. https://crsreports.congress.gov/product/pdf/R/R45206.

———. *U.S. Overseas Diplomatic Presence: Background and Issues for Congress*, by Cory Gill and Edward Collins-Chase. August 7, 2019. https://crsreports.congress.gov/product/pdf/IF/IF11242.

Vergun, David. "Acquisition Reform a Top DOD Priority." Department of Defense, February 5, 2019. https://www.defense.gov/Explore/News/Article/Article/1749362/acquisition-reform-a-top-dod-priority/.

Waller, Christopher J., and Lowell R. Ricketts. "Fed Balance Sheet: The Rise and (Eventual) Fall." Federal Reserve Bank of St. Louis Regional Economist, November 22, 2017. https://www.stlouisfed.org/publications/regional-economist/january-2014/the-rise-and-eventual-fall-in-the-feds-balance-sheet.

Warrick, Thomas, and Mark Massa. *The Future of DHS Project: Key Findings and Recommendations: Congressional Oversight.* Washington, DC: Atlantic Council, December 29, 2020. https://www.atlanticcouncil.org/in-depth-research-reports/issue-brief/the-future-of-dhs-project-congressional-oversight/.

———, and Caitlin Durkovich. *Future of DHS Project: Key Findings and Recommendations*, Washington, DC: Atlantic Council, August 2020. https://www.atlanticcouncil.org/content-series/future-of-dhs/future-of-dhs-project-final-report/.

Watts, Barry. *The US Defense Industrial Base: Past, Present and Future.* Washington, DC: Center for Strategic and Budgetary Assessmemts, October 15, 2008. https://csbaonline.org/research/publications/the-us-defense-industrial-base-past-present-and-future.

Weigley, Russel F. *The American Way of War: A History of United States Military Strategy and Policy.* Indianapolis, IN: Indiana University Press, 1977.

Whitaker, Alan G., Shannon A. Brown, Frederick C. Smith, and Elizabeth McKune. *The National Security Policy Process: The National Security Council and The Interagency System (2011)*. Washington, DC: Industrial College of the Armed Forces, 2011. https://issat.dcaf.ch/download/1761 9/205945/icaf-nsc-policy-process-report-08-2011.pdfThe%20National %20Security%20Policy%20Process:%20The%20National%20Security%2 0Council%20and%20Interagency%20System.

Whoriskey, Peter, Douglas MacMillan, and Jonathan O'Connell. "Doomed to Fail: Why a $4 Trillion Bailout Couldn't Revive the American Economy." *The Washington Post*, October 5, 2020. https://www. washingtonpost.com/graphics/2020/business/coronavirus-bailout-spending/.

Wilson, George C. *This War Really Matters: Inside the Fight for Defense Dollars*. Washington, DC: CQ Press, 2000.

Wilson, James Q. *Bureaucracy*. New York, NY: Basic Books, 1989.

Wlezien, Christopher. "The Public as Thermostat: Dynamics of Preferences for Spending." *American Journal of Political Science* 39, no. 4 (1995): 981-1000.

Wolf, Chad F. *FY 2021 Budget in Brief*. Department of Homeland Security, August 2, 2020. https://www.dhs.gov/publication/fy-2021-budget-brief.

Wolf, Martin. "Containing China Is Not a Feasible Option." *Financial Times*, February 2, 2021. https://www.ft.com/content/83a521c0-6abb-4efa-be48-89ecb52c8d01.

World Bank. "GDP per Capita (Current US$)—United States." https:// data.worldbank.org/indicator/NY.GDP.PCAP.CD?locations=US.

"Xi Jinping Is Trying to Remake the Chinese Economy." *The Economist*, August 15, 2020. https://www.economist.com/briefing/2020/08/15/xi-jinping-is-trying-to-remake-the-chinese-economy.

Zelizer, Julian. "'It's the Economy, Stupid' All over Again." *CNN*, May 8, 2020. https://www.cnn.com/2020/05/08/opinions/economy-2020-election-trump-biden-zelizer/index.html.

Zhu, Eric, and Thomas Orlik, "When Will China Rule the World? Maybe Never," Bloomberg, July 5, 2021, at https://www.bloomberg.com/

news/features/2021-07-05/when-will-china-s-economy-beat-the-u-s-to-become-no-1-why-it-may-never-happen?sref=nGJRPMyD.

Zweig, Martin E. *Martin Zweig's Winning on Wall Street.* New York, NY: Warner Books, 1997.

INDEX

ABOUT THE EDITORS

Susan Bryant is the Executive Director of Strategic Education International. She is also an adjunct professor at Georgetown University, a visiting lecturer at Johns Hopkins University, and a visiting research fellow at National Defense University. Dr. Bryant is a retired army colonel whose overseas assignments include Afghanistan, Jerusalem, and South Korea. She holds a doctorate from Georgetown University as well as masters' degrees from Yale University and the Marine Corps University. Her publications include *Military Strategy in the 21st Century* and *Finding Ender: Exploring the Intersections of Creativity, Innovation and Talent Management in the United States Armed Forces.*

Mark Troutman is an educator, consultant, and retired colonel. He is Chief Operating Officer of Strategic Education International and teaches business and national security economics at Georgetown University, Johns Hopkins University, University of Maryland, and George Mason University. Dr. Troutman is the former Director of the Center for Infrastructure Protection (George Mason University) and former Dean of the Eisenhower School (National Defense University). His military career included assignments in Europe, the Middle East, Asia, and the United States. A SAMS and US Army War College graduate, he also holds a PhD in Economics from George Mason University and a Master of Public Policy degree from Harvard University.

About the Contributors

Heidi Demarest is a lieutenant colonel in the US Army and an associate professor of American politics at the United States Military Academy. She is the Deputy Head of the Department of Social Sciences and holds a PhD from Harvard University and a BS from the United States Military Academy. She is the author of *US Defense Budget Outcomes: Volatility and Predictability in Army Weapons Funding* and *coeditor of US National Security Reform: Reassessing the National Security Act of 1947.*

John Ferrari is a Visiting Fellow at the American Enterprise Institute and retired from the US Army as a major general after 32 years of service, of which over a dozen years were associated with defense resourcing. He served as the US Army's top strategic resource manager as the Director of Program Analysis and Evaluation, in addition to time on the Joint Staff and at the Office of Management and Budget. Ferrari holds an MBA from the Wharton School and a BS in Computer Science from the United States Military Academy.

Jason Galui is Founder & CEO of 4 Liberty Consulting LLC, which collaborates at the nexus of business, government, and academia. He is a Professor of Practice at SMU Cox School of Business, a visiting scholar at Texas A&M, a Professor of Practice at JSOU, USSOCOM, and an adjunct professor at William & Mary. During his US Army career, Galui served on the NSC staffs for Presidents Obama and Trump, on the Council of Economic Advisers, as Special Assistant to the 18th Chairman of the Joint Chiefs of Staff, as an assistant professor at the United States Military Academy, and was deployed to Iraq and Afghanistan.

Michael Linick is a Senior Defense and International Policy Researcher at RAND and an adjunct assistant professor in the Security Studies Program at Georgetown University. A retired US Army colonel, he holds master's degrees from Georgetown University (national security), Catholic University (world politics), and the US Army War College (strategic studies). At RAND, his publications include *The Evolution of US Military Policy from the Constitution to the Present, The US Department of Defense's Planning Process: Components and Challenges*, and the game *Hedgemony: A Game of Strategic Choices*.

Thomas L. McNaugher retired from the position of Senior Visiting Professor in the Security Studies Program of Georgetown University's School of Foreign Service in June 2018 but continues to teach courses there on the defense budget and the American Civil War. He began teaching at Georgetown University in 2011 after long stints at the RAND Corporation and the Brookings Institution. Dr. McNaugher's books include *Arms and Oil: US Military Strategy and the Persian Gulf* and *New Weapons, Old Politics: America's Military Procurement Muddle*.

Geoffrey Odlum is a retired US State Department Foreign Service Officer who served in diplomatic assignments throughout Europe and the Middle East and in Washington, DC. He is a consultant on emerging technologies' impacts on national security, a Senior Advisor with the DOD-funded technology accelerator FedTech, and President of the Board of Coalescion, a nonprofit foreign and security assistance implementer. Odlum holds a graduate degree in national security studies from the National War College.

Rebecca D. Patterson is a Professor of Practice and Associate Director of the Security Studies Program at Georgetown University. A retired US Army lieutenant colonel, Dr. Patterson has served as the Deputy Director in the Office of Peace Operations, Sanctions, and Counter Terrorism at the Department of State. She holds a PhD from The George Washington University and is a lifetime member of the Council on Foreign Relations. She is the author of *The Challenge of Nation-Building: Implementing Effective Innovation in the US Army from World War II to the Iraq War*.

Laura Junor Pulzone is the Director of the Institute for National Strategic Studies at the National Defense University. She holds a PhD in economics from George Mason University. Dr. Pulzone has written and testified on topics ranging from military readiness to civilian personnel policy. She has held various executive leadership titles including the Deputy Undersecretary of Defense for Personnel and Readiness.

Cambria Rapid Communications in Conflict and Security (RCCS) Series

General Editor: Geoffrey R. H. Burn

The aim of the RCCS series is to provide policy makers, practitioners, analysts, and academics with in-depth analysis of fast-moving topics that require urgent yet informed debate. Since its launch in October 2015, the RCCS series has the following book publications:

- *A New Strategy for Complex Warfare: Combined Effects in East Asia* by Thomas A. Drohan
- *US National Security: New Threats, Old Realities* by Paul R. Viotti
- *Security Forces in African States: Cases and Assessment* edited by Paul Shemella and Nicholas Tomb
- *Trust and Distrust in Sino-American Relations: Challenge and Opportunity* by Steve Chan
- *The Gathering Pacific Storm: Emerging US-China Strategic Competition in Defense Technological and Industrial Development* edited by Tai Ming Cheung and Thomas G. Mahnken
- *Military Strategy for the 21st Century: People, Connectivity, and Competition* by Charles Cleveland, Benjamin Jensen, Susan Bryant, and Arnel David
- *Ensuring National Government Stability After US Counterinsurgency Operations: The Critical Measure of Success* by Dallas E. Shaw Jr.
- *Reassessing U.S. Nuclear Strategy* by David W. Kearn, Jr.
- *Deglobalization and International Security* by T. X. Hammes
- *American Foreign Policy and National Security* by Paul R. Viotti

- *Make America First Again: Grand Strategy Analysis and the Trump Administration* by Jacob Shively
- *Learning from Russia's Recent Wars: Why, Where, and When Russia Might Strike Next* by Neal G. Jesse
- *Restoring Thucydides: Testing Familiar Lessons and Deriving New Ones* by Andrew R. Novo and Jay M. Parker
- *Net Assessment and Military Strategy: Retrospective and Prospective Essays* edited by Thomas G. Mahnken, with an introduction by Andrew W. Marshall
- *Deterrence by Denial: Theory and Practice* edited by Alex S. Wilner and Andreas Wenger
- *Negotiating the New START Treaty* by Rose Gottemoeller
- *Party, Politics, and the Post-9/11 Army* by Heidi A. Urben
- *Resourcing the National Security Enterprise: Connecting the Ends and Means of US National Security* edited by Susan Bryant and Mark Troutman

For more information, visit www.cambriapress.com.

CPSIA information can be obtained
at www.ICGtesting.com
Printed in the USA
BVHW061609030522
635996BV00033B/2472